WITHDRAWN
UTSA LIBRARIES

D0793836

Bandwidth Extension of Speech Signals

Lecture Notes in Electrical Engineering

(*continues after index*)

Bernd Iser • Wolfgang Minker • Gerhard Schmidt

Bandwidth Extension
of Speech Signals

 Springer

Library
University of Texas
at San Antonio

Bernd Iser
Harman/Becker
Automotive Systems
Söflinger Str. 100
89077 Ulm
Germany
BIser@harmanbecker.com

Gerhard Schmidt
Harman/Becker
Automotive Systems
Söflinger Str. 100
89077 Ulm
Germany
GeSchmidt@harmanbecker.com

Wolfgang Minker
Institute of Information Technology
University of Ulm
Albert-Einstein-Allee 43
89081 Ulm
Germany
wolfgang.minker@uni-ulm.de

ISSN: 1876-1100   e-ISSN: 1876-1119
ISBN: 978-0-387-68898-5 e-ISBN: 978-0-387-68899-2
DOI: 10.1007/978-0-387-68899-2

Library of Congress Control Number: 2008927730

© 2008 Springer Science+Business Media, LLC
All rights reserved. This work may not be translated or copied in whole or in part without the written
permission of the publisher (Springer Science+Business Media, LLC, 233 Spring Street, New York, NY
10013, USA), except for brief excerpts in connection with reviews or scholarly analysis. Use in connection
with any form of information storage and retrieval, electronic adaptation, computer software, or by similar
or dissimilar methodology now known or hereafter developed is forbidden.
The use in this publication of trade names, trademarks, service marks, and similar terms, even if they are
not identified as such, is not to be taken as an expression of opinion as to whether or not they are subject
to proprietary rights.

Printed on acid-free paper

springer.com

Library
University of Texas
at San Antonio

Preface

The idea for this book was formed during the doctorate of Bernd Iser. Bernd Iser was working on efficient and robust bandwidth extension algorithms in hands-free systems for Harman/Becker Automotive Systems. It turned out that bandwidth extension of speech signals was a topic of appreciable interest, where lots of scientific publications discussing details of specific solutions could be found. What was missing was a contribution elaborately discussing the entirety of different approaches and comparing, respectively evaluating them in a meaningful manner. Another property that was disregarded in the state of the art was the influence of noise corrupted real-world signals. All these considerations led to the belief that there was a need for a book taking all these missing aspects into account.

Prof. Dr. Wolfgang Minker, who was supervising the doctorate of Bernd Iser at the University of Ulm, Germany, came up with the idea of writing such a book. Dr. Gerhard Schmidt, the leader of the acoustic signal processing research team, Bernd Iser was working at Harman/Becker Automotive Systems, joined the project for having another speech signal processing expert on board. Based on the research work on the topic of bandwidth extension of speech signals the present book emerged in collaboration that tries to cover the above described requirements.

Like all extensive projects this book project would not have been possible to handle without the support, advice, criticism and review of countless people. Representative for all of them a few people that added substantial contributions to this book should be mentioned and thanked for their valuable support.

The authors thank Dr. Markus Buck for his accurate and detailed review of the manuscript; Mohamed Krini for many valuable discussions on the algorithms as well as on the manuscript and Dr. Tim Haulick for arranging the possibility of conducting research on the topic of bandwidth extension of speech signals at Harman/Becker Automotive Systems.

Bernd Iser
Wolfgang Minker
Gerhard Schmidt

Contents

1

Introduction

Speech is a natural and therefore privileged communication modality. This is the reason for the great success of speech driven services and speech based media. Multimedia would not be imaginable without high quality audio. Today's environment is full of high quality audio sources like CD-Audio, DVD-Audio, radio broadcast, television broadcast, and so on.

One of the oldest but still most popular media based on audio is the telephone network. But since its invention in the nineteenth century capabilities and simultaneously demands on audio quality have increased [Bell 77]. Today's telephone networks still provide poor audio quality due to historical limitations (see Sect. 1.3).

The reason for the poor audio quality of the telephone network is the very limited bandwidth that is provided. Analog networks, for example, provide only a bandwidth of about 3.1 kHz [ITU 88b] (see Fig. 1.1b). This leads to reduced speech quality and even intelligibility. A typical property of analog networks is the difficulty of distinguishing between several fricatives like present in the words "feel" ([fiːl][1]) and "veal" ([viːl]). Another typical problem is that one is not able to distinguish similar voices (father–son problem) over the telephone.

By using more modern media this drawback becomes even more obvious. This can be best experienced by listening to radio or CD within a car and afterwards using the hands-free system. Another example are the telephone services that are recently available in the internet like skype^{TM}, which have a considerably higher bandwidth than conventional telephone networks. However the telephone network is still one of the most widespread networks all over the world. This deed has made attempts to change the network in order to provide a better audio quality doomed to failure due to the massive effort of exchanging the hardware.

---

[1] Phonetic description according to [IPA 49].

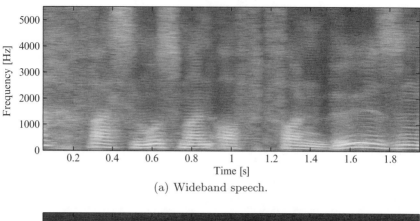

(a) Wideband speech.

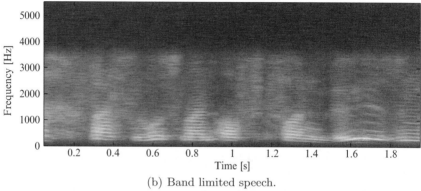

(b) Band limited speech.

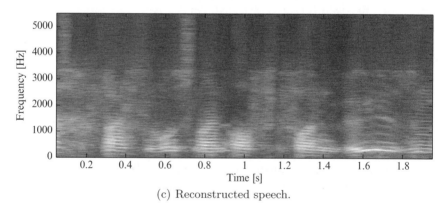

(c) Reconstructed speech.

Fig. 1.1. Spectrograms of (**a**) wideband speech, (**b**) band limited speech, and (**c**) reconstructed speech. The spectrograms (**a**) and (**c**) are limited by the sampling rate of 11,025 Hz. The spectrogram in (**b**) is limited by an analog telephone bandpass with a passband from approximately 300 to 3,400 Hz causing a reduced speech quality in comparison to (**a**) which could be a recording close to the mouth with no attenuation. The spectrogram (**c**) contains the estimates for the frequency components missing in (**b**). Methods for estimating these components will be subject of this work

This is the point where the idea of bandwidth extension comes into mind quite intuitively. Bandwidth extension in this context means the estimation of the not transmitted frequency components out of the transmitted signal by exploiting the transinformation included in speech signals and therewith increasing the speech quality (see Fig. 1.1c). This approach yields the advantage that nothing has to be changed within the network – it is simply an optional feature on the terminal side.

1.1 Scope of the Book

This book is intended to provide a profound knowledge on the problem of bandwidth extension of band limited speech signals in a first step. Theory and methods for quality enhancement of speech signals that have undergone a bandlimitation, as it is the case for example in a telephone network, will be described. These enhancements have been performed with and without the presence of noise. In a second step, novel, not yet published solutions as well as improvements for the problem of bandwidth extension will be presented. In contrast to prior work emphasis is placed on novel approaches to handle problems that emerge from dealing with real-world signals and environments comprising all kinds of disturbances. Another previously not considered aspect is the subjective evaluation and comparison of the different methods. By analyzing the usability of well known objective distance measures the need of a measure that better correlates with subjective evaluation resulted in the development of a novel objective distance measure.

All band limited signals that are matter of interest in this book are speech signals that have undergone a band limitation by the application of a telephone bandpass. But the presented algorithms are not restricted to telephone band limited signals. They are also applicable to speech signals that are band limited in any other way as for example by downsampling with prior lowpass filtering. Another field of application would be the re-synthesis of heavily perturbed speech signals (see [Hosoki 02, Seltzer 05b, Yegnanarayana 97]) that can not sufficiently be enhanced in terms of quality by the standard noise reduction systems [Hänsler 03]. The scope of the book will include topics related to speech coding, pattern and speech recognition, speech enhancement, statistics and digital signal processing in general.

The study and development of several methods for the bandwidth extension of speech signals will be described. Problems and the respective solutions are discussed for the different approaches. Since all described methods will be based on the source-filter model, this model will be presented after a short problem motivation and illustration of the process of human speech generation. The proposed methods will include the extraction of speech-model based parameters like cepstral coefficients, auto-regressive parameters (all-pole-filter coefficients), line spectral pairs (LSP) or mel-frequency cepstral coefficients

(MFCC). These parameters and their properties will briefly be presented. Besides these parametric representations of the spectral envelope some other scalar speech features like the zero-crossing rate, gradient index, pitch frequency, local kurtosis, spectral centroid, and energy based features will be introduced. For evaluating the different approaches as well as for the training and operation, a couple of well known spectral distortion measures and a new measure will be presented. After these rather general topics the main algorithms will be introduced. These can be divided into two separate subtasks, namely the generation of a broadband excitation signal and the generation of a broadband spectral envelope. For the first one, several approaches like non-linear characteristics, spectral shifting, signal generators, etc. will be explained. For the latter one, codebook approaches, neural networks, linear mapping, as well as joint approaches will be discussed. For the generation of the broadband spectral envelope the proposed methods need a prior training phase. This part and the speech corpora which have been used as well as the specific demands and preparation of these corpora will be part of this section. Another aspect of this section will be the challenge of facing different transmission paths with their respective frequency responses that have not been part of the training scenario and schemes to overcome this problem. Afterwards focus will be placed on the performance of these methods. For the evaluation of the performance the already introduced objective spectral distortion measure will be used. Additionally, a subjective evaluation will be presented by analyzing the result of listeners voting the different approaches. A significance analysis will also be drawn for the subjective evaluation as well as a rank correlation between the subjective results and the respective objective distortion measures to evaluate the significance of these objective measures. The most promising method has been implemented in a real-time environment using appropriate hardware. This system will be described as well. The book will close with a summary of the described methods and the results of the evaluation.

The speech signals within this book emerge from the car environment and have a sampling rate of 11,025 Hz and a resolution of 16 bit. This is the sampling rate and the resolution that would be available for example when using one subsampled (factor 4) channel of a MOST-bus (Multimedia Oriented System Transfer) as it is installed in many present cars. However, the presented algorithms are neither restricted to this sampling rate nor to the car environment.

1.2 Nature of Speech Signals

Human speech generally occupies approximatively the whole frequency range that is perceptible for the auditory system [Zwicker 99]. In Fig. 1.2 the estimate for the power spectral density of a speech sample recorded with a sampling rate of 44.1 kHz is depicted. It is clearly visible that this signal possesses

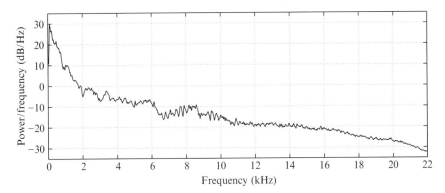

Fig. 1.2. Power spectral density estimate of a speech sample (male speaker; approximately 40 s of speech). Note that the signal has not been equalized for the microphone transfer function

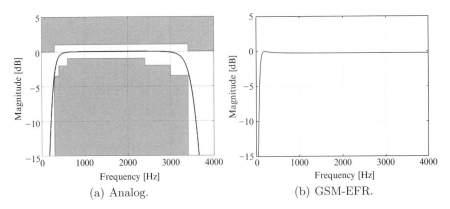

Fig. 1.3. (a) Telephone bandpass according to [ITU 88b] and (b) highpass as implemented in GSM-EFR according to [ETSI 00]

power up to 22.05 kHz. This broad frequency range however is not necessary for speech intelligibility. The speech intelligibility decreases only marginally if band limiting the speech signal by using a sampling rate of 16 kHz for example. However, the speech quality decreases remarkably.

1.3 Band Limited Speech Signals

The degradation of speech quality using analog telephone systems is caused by the introduction of band-limiting filters within amplifiers used to keep a certain signal level in long local loops [Kammeyer 92]. These filters have a passband from approximately 300 up to 3,400 Hz (see Fig. 1.3a) and are applied to reduce the crosstalk between different channels.

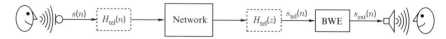

Fig. 1.4. Overall system for bandwidth extension. The *dashed line* indicates that the location(s) where the band limitation(s) take(s) place depend(s) on the routing of the call

As can be seen in Fig. 1.1b the application of such a bandpass (depicted in Fig. 1.3a) attenuates considerable speech portions. Digital networks, such as the integrated service digital network (ISDN) and the global system for mobile communication (GSM), are able to transmit speech in higher quality since signal components below 300 Hz as well as components between 3.4 and 4 kHz can be transmitted (see Fig. 1.3b). Considering a transmission over a GSM network the only two bandlimiting factors are the sampling rate of 8 kHz limiting the signal to an upper limit of 4 kHz and the lowpass filter depicted in Fig. 1.3b that is specified for example in the enhanced full-rate coder [ITU 88b]. However, this is only true if the entire call (in terms of its routing) remains in those networks – when leaving into an analog telephone network, the speech signal is once again band limited (see dashed rectangles in Fig. 1.4).

Thus, great efforts have been made to increase the quality of telephone speech signals in recent years. Wideband codecs are able to increase the bandwidth up to 7 kHz or even higher at only moderate complexity [ITU 88a, Croll 72]. Other approaches try to increase the bandwidth by transmitting side information on the missing frequency bands or by combined estimation and coding (see [Agiomyrgiannakis 04, Geiser 05]). Nevertheless, applying these codecs would require an exchange of the current networks or at least, in the second case, the usage of two devices that are able to encode side information using acoustic watermarking on the far end side and to decode such information on the local side. Acoustic watermarking in this context means to hide information in the audio data that is however not audible but still possible to decode. Another possibility is to increase the bandwidth after transmission by means of bandwidth extension (for other approaches, e.g. psychoacoustic approaches, see [Boillot 05]). The basic idea of these enhancements is to estimate the speech signal components above 3400 Hz and below 300 Hz and to complement the signal in the idle frequency bands with this estimate. Precondition is the sufficient correlation between the speech signal in the telephone band and the extension regions which is discussed in [Jax 02b]. In this case the telephone networks remain untouched. Figure 1.4 shows the basic structure of a bandwidth extension (BWE) system included in the receiving path of a telephone connection.

Additionally, three time-frequency analyses are presented in Fig. 1.1. The first analysis depicts a wideband speech signal $s(n)$ recorded close to the mouth of the communication partner on the remote side. If we assume no errors or distortions during the transmission, a bandlimited signal $s_{tel}(n)$

as depicted in the center diagram would be received at the local side. The truncation of the frequency range is clearly visible. Without any additional processing the local communication partner would be listening to this signal. If bandwidth extension is applied a signal $s_{\text{ext}}(n)$ as depicted in part (c) of Fig. 1.1 would be reconstructed. Even if the signal is not exactly the same as the original one, it sounds more natural and – as a variety of listening test indicate – the speech quality in general is increased as well [Iser 03].

1.4 Organization of the Book

In Chap. 2 the human speech generation process will be introduced briefly. Based on a mechanical abstraction a model for this speech production process will be presented. As mentioned before this model can be split into two independent parts, namely the generation of an excitation signal and the spectral coloration of this white signal by an all-pole filter representing the resonance behavior of the vocal tract. After the introduction of this model a basic scheme for bandwidth extension exploiting the model for the speech generation process will be presented. All discussed algorithms within this book will make use of this basic structure.

In Chap. 3 the required speech analysis techniques will be introduced. These techniques include the so-called *linear predictive analysis* as well as other vectorial and scalar parametric representations of the spectral envelope. The chapter will close with the introduction of several distance measures that will later on be needed to evaluate different estimation methods as well as for training purposes and during the operation of such a system.

In Chap. 4 the extraction of the narrowband excitation signal as well as several methods for the generation of a broadband excitation signal will be discussed. Another issue of this chapter will be the power adjustment that is needed for the estimated broadband excitation signal. The chapter will close with a brief discussion of the presented methods.

In Chap. 5 the estimation of the broadband spectral envelope as well as the different methods for this task will be the main topic. Another topic will be the preparation of the speech-data base for the training phase of the respective methods. Analogous to the previous chapter this chapter will close with a brief discussion of the presented methods.

In Chap. 6 subjective as well as objective quality criteria to evaluate the different methods will be presented. This includes the introduction of a novel objective quality criterion. A significance analysis of the subjective results will be drawn. Additionally the rank correlation between the objective results and the subjective ones will be computed to evaluate the objective measures.

In Chap. 7 a summary of the whole achieved results will be drawn.

2

Speech Generation

In the following a cursory description of the human speech generation process will be given. To emulate the speech generation process it is mandatory to get a profound understanding of the mechanical processes within the involved organs. Special emphasis is placed on the description of an appropriate model for the human speech generation process. The derivation of an adequate model will result in the proposal of an intuitive approach for bandwidth extension of band limited speech signals.

2.1 Human Speech Generation Process

In Fig. 2.1 a mechanical abstraction of the human speech generation process is depicted [Korpiun 02]. By breathing, the lungs get filled with air. This air being expelled through the trachea causes the tensed vocal cords (see Fig. 2.2) within the larynx tube to vibrate. The resulting air flow through the opening of the vocal cords, the so-called *glottis*, gets chopped into periodic pulses by the periodically opening and closing vocal cords. This is the case when voiced sounds like [ə][1], [æ], [ɑː], [ɛ], [ɪ], or [iː] are being produced. The inverse of the corresponding time period is called fundamental frequency or pitch frequency and ranges from about 80 Hz for male speakers up to 300 Hz for female speakers or children. In the case of unvoiced sounds like the fricatives [ʃ], [s], [θ], or [f] the vocal cords are loose causing a turbulent, noise-like air flow [Rabiner 93].

In Fig. 2.3 the glottal sound pressure difference that could be observed directly behind the vocal cords is depicted for a voiced and an unvoiced sound. In the case of the voiced utterance the periodicity is clearly visible. The period is 10 ms corresponding to a pitch frequency of 100 Hz. The recording was done with a male participant.

---

[1] Phonetic description according to [IPA 49].

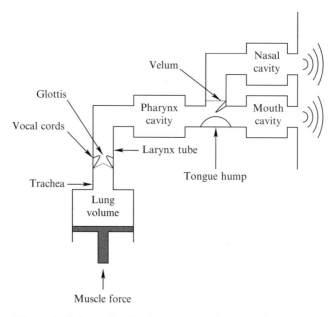

Fig. 2.1. Scheme for the human speech generation process

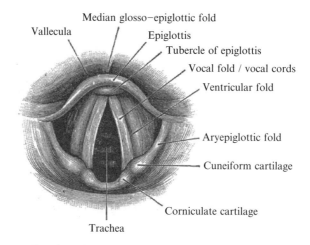

Fig. 2.2. Vocal cords and glottis in top view according to [Gray 18]

After passing the vocal cords the air flow reaches several cavities namely the pharynx cavity, the nasal cavity, and the mouth cavity which all together form the so-called *vocal tract*. The cavities within the vocal tract can additionally be separated from each other and changed in size and shape by the tongue, the velum, the jaw, the mouth, and the lips. In these cavities with their

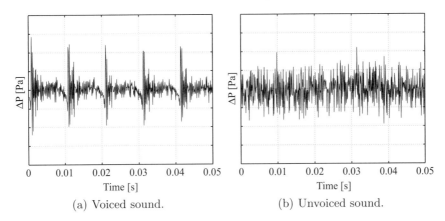

Fig. 2.3. Glottal air flow respectively the sound pressure difference for (**a**) a voiced sound and (**b**) an unvoiced sound

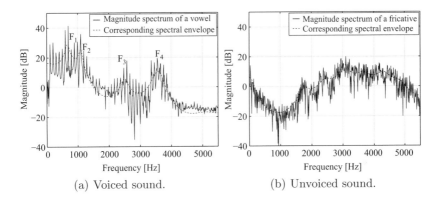

Fig. 2.4. Short-term spectrum of a (**a**) voiced utterance and (**b**) unvoiced utterance

special characteristics concerning resonance the final sounds are produced. The characteristic of distinct human voices is build up by the mixture of a pitch frequency which varies only marginally around a working point and the characteristics of the vocal tract [Eppinger 93, Deller Jr. 00, Rabiner 78, Rabiner 93]. In Fig. 2.4 the short-term spectra of a voiced and unvoiced utterance are depicted. Since speech is a quasi-stationary process the approach using these short-term spectra is feasible (in the following these short-term spectra will mostly be called spectra for reasons of brevity). One can see clearly the pitch structure in Fig. 2.4a with its maxima every 100 Hz. The local maxima of the envelope spectrum are called *formants* and are denoted with F_1 to F_4. The terms pitch and envelope spectrum will become clearer in the next section when a model for the speech production process is introduced.

2.2 Model for the Speech Generation Process

After a brief description of the human speech generation process we will now introduce the well known source-filter model [Deller Jr. 00] as an abstraction to this process. The source-filter model separates the speech production into two independent processes:

- The first process is the generation of a so-called *excitation signal* $e(n)$.
- The second process describes the influence of the vocal tract on this excitation signal to build up the final speech signal $s(n)$.

The source-filter model is depicted in Fig. 2.5. Left of the dashed line the source part of the model can be seen. Right of the dashed line the filter part is depicted. The parameters $f_0(n)$, $g(n)$, $\sigma(n)$ and $H(z,n)$ will be described step by step in the following sections.

2.2.1 Excitation Signal

As mentioned before, the excitation signal corresponds to the signal that could be observed directly behind the vocal cords. This part of the source-filter model is called the source part. For the generation of the excitation signal the source part of the source-filter model once again differentiates between two scenarios:

- For voiced sounds the excitation signal is modeled by a pulse train.
- For unvoiced sounds a noise generator models the excitation signal.

Depending on the pitch frequency $f_0(n)$ the periodicity of the pulse train can be adjusted. With $\sigma(n)$ the model gain is characterized. The amount of voicedness is controlled via the factor $g(n)$, with $0 \leq g(n) \leq 1$. The shape of the pulse train ideally corresponds to the shape of the glottal pulses. The resulting excitation signal is labeled with $e(n)$. In Fig. 2.6 the excitation signals for a voiced and an unvoiced utterance are depicted. It is obvious, that the signals have a flat short-term spectral envelope and in the case of the voiced utterance the periodicity or pitch, respectively, is clearly visible.

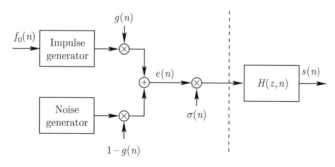

Fig. 2.5. Model for the human speech production process

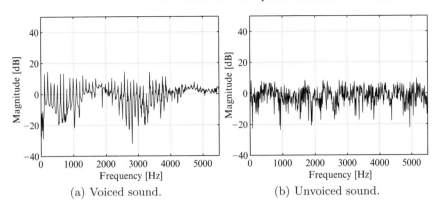

(a) Voiced sound.   (b) Unvoiced sound.

Fig. 2.6. Spectrum of the excitation signal for (**a**) a voiced utterance and (**b**) an unvoiced utterance

2.2.2 Vocal Tract Filter

The influence of the vocal tract on the excitation signal is modeled by a discrete time-variant filter $H(z, n)$. In this so-called filter part the influence of the pharynx-, nasal-, and mouth cavity on the excitation signal is emulated. First modeling approaches for this cavities using lossless tubes with a variable but discretized length and diameter (see Sect. 3.2.1) lead to the use of an autoregressive filter [Rabiner 78]. The structure of this filter is purely recursive leading to an all-pole filter

$$H(z, n) = \frac{1}{A(z, n)} = \frac{1}{1 - \sum_{k=1}^{P} a_k(n) z^{-k}}. \tag{2.1}$$

The coefficient with index zero is normalized to $a_0 = 1$. Hence, we can state

$$\frac{1}{2\pi} \int_{-\pi}^{\pi} \frac{1}{H\left(e^{j\Omega}, n\right)} \, d\Omega = \frac{1}{2\pi} \int_{-\pi}^{\pi} H\left(e^{j\Omega}, n\right) \, d\Omega = 1. \tag{2.2}$$

Thus, one property of the transfer function of this filter is the independence on the short-term power of the speech signal [Jax 02a]. Therefore it only represents the shape of the spectral envelope of the speech signal. In Fig. 2.7 the spectral envelopes of a voiced and of an unvoiced signal are depicted.

Another property of $A(z, n)$ is that it is a minimum phase filter and therefore its inverse $H(z, n)$ is guaranteed to be a stable filter [Markel 76]. This will be important in Sects. 3.1 and 4.1. As shown in [Markel 76] a third property of $A\left(e^{j\Omega}, n\right)$ is

$$\frac{1}{2\pi} \int_{-\pi}^{\pi} \ln\left|A\left(e^{j\Omega}, n\right)\right|^2 \, d\Omega = 0. \tag{2.3}$$

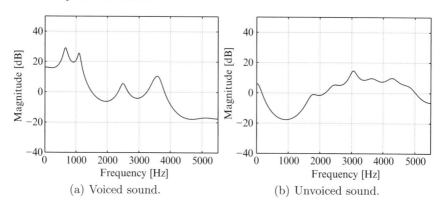

(a) Voiced sound.   (b) Unvoiced sound.

Fig. 2.7. Spectrum of the envelope for (**a**) a voiced utterance and (**b**) an unvoiced utterance

This means that the area of the log magnitude spectra of $1/A\left(e^{j\Omega},n\right)$ above $0\,\mathrm{dB}$ equals the area below $0\,\mathrm{dB}$.

With this transfer function we can express the speech signal as a combination of weighted previous speech samples and the excitation signal $e(n)$:

$$s(n) = \underbrace{\sum_{k=1}^{P} a_k(n)s(n-k)}_{\text{Filter part}} + \underbrace{\sigma(n)\,e(n)}_{\text{Source part}} \quad . \tag{2.4}$$

2.3 Basic Approach for Bandwidth Extension

Based on the source-filter model for the speech production process we will now introduce an intuitive approach for bandwidth extension algorithms [Iser 05b, Iser 05a, Kornagel 06]. The basic scheme for these algorithms is depicted in Fig. 2.8. Non-model based approaches can be found for example in [Croll 72, Hermansky 95, Laaksonen 05, Yasukawa 96a, Yasukawa 96b, Yasukawa 96c]. Starting from a band limited speech signal $s_{\mathrm{nb}}(n)$ and following the replication of the source path the narrowband excitation signal $e_{\mathrm{nb}}(n)$ is estimated. Afterwards the broadband excitation signal $\hat{e}_{\mathrm{bb}}(n)$ is generated out of the narrowband excitation signal. In the next step the spectrally flat estimation of the broadband excitation signal is colored by the application of the estimated broadband spectral envelope resulting from the coefficients

$$\hat{\mathbf{a}}_{\mathrm{bb}}(n) = [\hat{a}_0^{\mathrm{bb}}(n), \hat{a}_1^{\mathrm{bb}}(n), \ldots, \hat{a}_{N_{\mathrm{bb}}-1}^{\mathrm{bb}}(n)]^{\mathrm{T}}, \tag{2.5}$$

where N_{bb} represents the order of the broadband all-pole filter. These coefficients are estimated in the replication of the filter path that also starts from

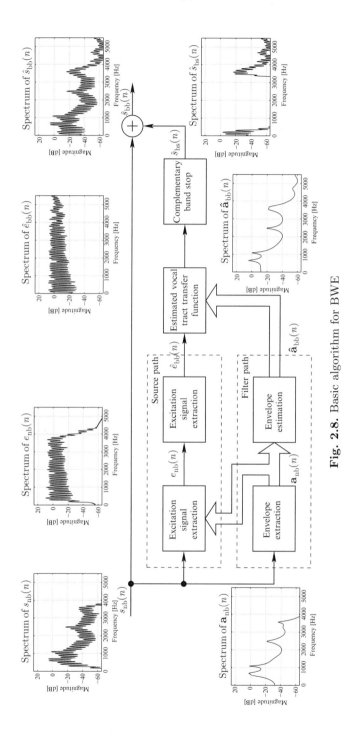

Fig. 2.8. Basic algorithm for BWE

the band limited speech signal $s_{nb}(n)$. Then the narrowband spectral envelope or the representing coefficients

$$\mathbf{a}_{nb}(n) = [a_0^{nb}(n), a_1^{nb}(n), \ldots, a_{N_{nb}-1}^{nb}(n)]^T \qquad (2.6)$$

respectively are estimated. Analogous to above N_{nb} denotes the narrowband all-pole filter order. These coefficients are useful for the above mentioned estimation of the narrowband excitation signal. Furthermore these coefficients are required for the estimation of the broadband spectral envelope $\hat{\mathbf{a}}_{bb}(n)$. After the coloration of the estimated broadband excitation signal $\hat{e}_{bb}(n)$ with the estimated broadband spectral envelope we have a completely synthesized speech signal. Since the original signal within the telephone band is present, a band stop filter is used to get rid of the redundant frequency components. Finally, the complementary frequency components are combined within the summation unit resulting in an artificially supplemented and thereby quality improved telephone speech signal.

The usage of the source-filter model in an approach for bandwidth extension of speech signals is motivated by its extensive use and success in the field of speech coding. The non-model based approaches mentioned at the beginning of this section ([Croll 72, Hermansky 95, Laaksonen 05, Yasukawa 96a, Yasukawa 96b, Yasukawa 96c]) do mostly have the advantage that they are very simple. The problem is that they depend exceedingly on the kind of transmission and the corresponding transfer function, respectively. In most cases the quality of these non-model based approaches is generally speaking rather fair and depends crucially on the sampling rate used. Other models for the speech production process will briefly be mentioned in the next chapter when discussing different analysis techniques. However these models lead to similar results.

3

Analysis Techniques for Speech Signals

In this chapter we will introduce several well known approaches for speech analysis. These approaches include parametric representations of the spectral envelope of a speech signal as well as scalar speech features. These features will later on be used for representing the spectral envelope for classification and extension schemes. The scalar features are interesting for more robust classification as they are able for example to give information on the gender of the speaker or on voicedness of the utterance. The presented methods are related to or originate from the field of speech coding or speech recognition [Mitra 93, Kleijn 95]. To evaluate speech coders some distance measures have been introduced that will also be presented and complemented by a novel measure. These distance measures will allow us to evaluate the quality of the bandwidth extension approaches that will be presented in Chap. 4 and Chap. 5. The focus of this chapter is placed on the linear predictive analysis of speech signals, which will result in a parametric representation of the spectral envelope, since all algorithms that are presented in this book make use of this method.

3.1 Linear Predictive Analysis

Figure 3.1 shows the linear model for the speech production process that has already been introduced in Sect. 2.2. The idea that forms the basis of the linear predictive analysis is the approach of approximating the current speech sample $s(n)$ by a linear combination of past values leading to

$$s(n) \approx a_1(n)s(n-1) + a_2(n)s(n-2) + \ldots + a_P(n)s(n-P). \qquad (3.1)$$

Here the coefficients $a_i(n)$ form the coefficients of a straight recursive filter of order P. By including an excitation term $\sigma(n)e(n)$ we can set this approximation equal to the actual speech sample

$$s(n) = \sum_{i=1}^{P} a_i(n)s(n-i) + \sigma(n)e(n), \qquad (3.2)$$

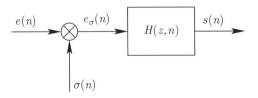

Fig. 3.1. Linear model for the speech production process

where $e(n)$ denotes a normalized excitation signal and $\sigma(n)$ its gain. By performing a transformation into the z-domain and considering the a_i as well as σ as constant within a certain window we get

$$S(z) = \sum_{i=1}^{P} a_i z^{-i} S(z) + \sigma E(z). \tag{3.3}$$

The transfer function of this system is given by

$$H(z) = \frac{S(z)}{\sigma E(z)}. \tag{3.4}$$

By solving (3.3) for $\sigma E(z)$ and inserting this result into (3.4) we can finally formulate the transfer function as

$$H(z) = \frac{1}{1 - \sum_{i=1}^{P} a_i z^{-i}} = \frac{1}{A(z)}. \tag{3.5}$$

The system $H(z)$ is obviously an all-pole system. Let us now recall (3.1) and (3.2) and formulate an estimate $\hat{s}(n)$ for the actual speech sample out of a linear combination of previous ones

$$\hat{s}(n) = \sum_{i=1}^{P} a_i s(n-i). \tag{3.6}$$

With this estimate we can now define the prediction error

$$e_\sigma(n) = \sigma e(n), \tag{3.7}$$

according to Fig. 3.2 as

$$e_\sigma(n) = s(n) - \hat{s}(n) = s(n) - \sum_{i=1}^{P} a_i s(n-i). \tag{3.8}$$

With the respective formulation in the z-domain

$$E_\sigma(z) = S(z) - \sum_{i=1}^{P} a_i z^{-i} S(z) = S(z) \left[1 - \sum_{i=1}^{P} a_i z^{-i} \right], \tag{3.9}$$

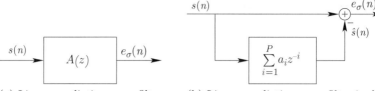

(a) Linear prediction error filter in symbolic view.

(b) Linear prediction error filter in detailed view.

Fig. 3.2. (a) and **(b)** Show a predictor-error filter

we can set up the corresponding error transfer function

$$A(z) = \frac{E_\sigma(z)}{S(z)} = 1 - \sum_{i=1}^{P} a_i z^{-i}. \tag{3.10}$$

Due to the fact that speech is a time variant process the predictor coefficients have to be estimated periodically for a short, quasi-stationary, speech interval starting at a given time n. For this reason let us initially define a short segment of speech as well as a short segment of the predictor error signal which will be specified in more detail in the respective sections later on

$$s_n(m) = s(n + m), \tag{3.11}$$

$$e_{n,\sigma}(m) = e_\sigma(n + m). \tag{3.12}$$

To achieve a set of optimal predictor coefficients $a_{i,\text{opt}}$ (optimal in the sense of a minimum mean squared error approach) for the current speech segment let us first define the sum of the squared error of the predictor error signal within this speech segment

$$E_n = \sum_m e_{n,\sigma}^2(m). \tag{3.13}$$

According to (3.8) this can be written as

$$E_n = \sum_m \left[s_n(m) - \sum_{i=1}^{P} a_i s_n(m - i) \right]^2. \tag{3.14}$$

Since we are looking for the minimum of the mean squared error depending on the optimal predictor coefficients $a_{i,\text{opt}}$ we now differentiate E_n with respect to the a_i and set the result equal to zero

$$\left. \frac{\partial E_n}{\partial a_i} \right|_{a_i = a_{i,\text{opt}}} = 0, \qquad \text{for } i \in \{1, .., P\}. \tag{3.15}$$

Using (3.14) this can be written as

$$\frac{\partial}{\partial a_\ell} \sum_m \left[s_n(m) - \sum_{i=1}^{P} a_i s_n(m-i) \right]^2 \Bigg|_{a_\ell = a_{\ell,\text{opt}}} = 0. \qquad (3.16)$$

By applying the chain rule we can formulate (in the following we assume $a_i = a_{i,\text{opt}}$)

$$0 = \sum_m 2 \left[s_n(m) - \sum_{i=1}^{P} a_i s_n(m-i) \right] \cdot \frac{\partial}{\partial a_\ell} \left[s_n(m) - \sum_{i=1}^{P} a_i s_n(m-i) \right]$$

$$0 = \sum_m 2 \left[s_n(m) - \sum_{i=1}^{P} a_i s_n(m-i) \right] \cdot [-s_n(m-\ell)]$$

$$0 = \sum_m -s_n(m)s_n(m-\ell) + s_n(m-\ell) \sum_{i=1}^{P} a_i s_n(m-i), \qquad (3.17)$$

for $\ell \in \{1, \ldots, P\}$. By further reorganization and following permutation of the sums we can write

$$\sum_m s_n(m)s_n(m-\ell) = \sum_m \sum_{i=1}^{P} a_i s_n(m-i)s_n(m-\ell),$$

$$\sum_m s_n(m)s_n(m-\ell) = \sum_{i=1}^{P} a_i \sum_m s_n(m-i)s_n(m-\ell), \qquad (3.18)$$

for $\ell \in \{1, \ldots, P\}$. If we now define the short-term estimate of the covariance function[1] of a signal $s_n(m)$ as

$$\phi_{ss,n}(i,\ell) = \sum_m s_n(m-i)s_n(m-\ell) \qquad (3.19)$$

we can write (3.18) as

$$\phi_{ss,n}(0,\ell) = \sum_{i=1}^{P} a_i \phi_{ss,n}(i,\ell), \text{ for } \ell \in \{1, \ldots, P\}. \qquad (3.20)$$

This forms a set of P equations in P unknowns which can already be solved. Using (3.14) as well as (3.19) and (3.20) we can now express the minimum mean squared error as

---

[1] Note that the notation is accurate only for zero mean processes.

$$
\begin{aligned}
E_{n,\text{error}} &= \sum_m \left[s_n(m) - \sum_{i=1}^{P} a_i s_n(m-i) \right]^2 \\[2mm]
&= \sum_m \left[s_n^2(m) - 2 s_n(m) \sum_{i=1}^{P} a_i s_n(m-i) \right. \\[2mm]
&\quad \left. + \sum_{i=1}^{P} a_i s_n(m-i) \sum_{\ell=1}^{P} a_\ell s_n(m-\ell) \right] \\[2mm]
&= \sum_m s_n^2(m) - 2 \sum_m s_n(m) \sum_{i=1}^{P} a_i s_n(m-i) \\[2mm]
&\quad + \sum_m \sum_{i=1}^{P} a_i s_n(m-i) \sum_{\ell=1}^{P} a_\ell s_n(m-\ell) \\[2mm]
&= \sum_m s_n^2(m) - 2 \sum_{i=1}^{P} a_i \sum_m s_n(m) s_n(m-i) \\[2mm]
&\quad + \sum_m \sum_{i=1}^{P} a_i s_n(m-i) \sum_{\ell=1}^{P} a_\ell s_n(m-\ell) \\[2mm]
&= \sum_m s_n^2(m) - 2 \sum_{i=1}^{P} a_i \phi_{ss,n}(i,0) + \sum_m \sum_{i=1}^{P} a_i s_n(m-i) \sum_{\ell=1}^{P} a_\ell s_n(m-\ell) \\[2mm]
&= \sum_m s_n^2(m) - 2 \sum_{i=1}^{P} a_i \phi_{ss,n}(i,0) + \sum_{i=1}^{P} a_i \sum_{\ell=1}^{P} a_\ell \sum_m s_n(m-i) s_n(m-\ell) \\[2mm]
&= \sum_m s_n^2(m) - 2 \sum_{i=1}^{P} a_i \phi_{ss,n}(i,0) + \sum_{i=1}^{P} a_i \sum_{\ell=1}^{P} a_\ell \phi_{ss,n}(i,\ell) \\[2mm]
&\overset{(3.20)}{=} \sum_m s_n^2(m) - 2 \sum_{i=1}^{P} a_i \phi_{ss,n}(i,0) + \sum_{i=1}^{P} a_i \phi_{ss,n}(i,0) \\[2mm]
&= \sum_m s_n^2(m) - \sum_{i=1}^{P} a_i \phi_{ss,n}(i,0) \\[2mm]
&= \phi_{ss,n}(0,0) - \sum_{i=1}^{P} a_i \phi_{ss,n}(i,0). \tag{3.21}
\end{aligned}
$$

Until now we have not yet specified the precise range m of speech that is used as well as the range over which the mean-squared error is computed. In the following two sections we will present two common methods of defining these ranges and the resulting equations that have to be solved [Rabiner 93].

3.1.1 Autocorrelation Method

A quite intuitive approach of defining the range of speech used within the summations is to set the speech signal outside the interval $m \in \{0, \ldots, N-1\}$ to zero. This can be done using a length N window function. Representative for a whole bunch of possible window functions we will use a rectangular window resulting in the definition of a speech interval

$$s_n(m) = \begin{cases} s(n+m), & \text{for } m \in \{0, \ldots, N-1\}, \\ 0, & \text{otherwise.} \end{cases} \tag{3.22}$$

This definition of a speech interval involves the definition of a segment of the squared error

$$E_n = \sum_{m=0}^{N-1+P} e_{n,\sigma}^2(m), \tag{3.23}$$

that has to be computed, leading to (see (3.19))

$$\phi_{ss,n}(i, \ell) = \sum_{m=0}^{N-1+P} s_n(m-i)s_n(m-\ell) \tag{3.24}$$

$$= \sum_{m=0}^{N-1-(i-\ell)} s_n(m)s_n(m+i-\ell). \tag{3.25}$$

Since (3.25) is only a function of $i - \ell$, the covariance $\phi_{ss,n}(i, \ell)$ can now be replaced by the autocorrelation function $r_{ss,n}(i - \ell)$ [Rabiner 93]

$$\phi_{ss,n}(i, \ell) = \sum_{m=0}^{N-1-(i-\ell)} s_n(m)s_n(m+i-\ell) = r_{ss,n}(i-\ell), \tag{3.26}$$

where the autocorrelation function is defined as

$$r_{ss,n}(i) = \sum_{m=0}^{N-1-i} s_n(m)s_n(m+i). \tag{3.27}$$

Considering the symmetry of the autocorrelation function we can now write (3.20) as

$$r_{ss,n}(l) = \sum_{i=1}^{P} a_i r_{ss,n}(|i - \ell|), \text{ for } \ell \in \{1, \ldots, P\}. \tag{3.28}$$

In matrix-vector notation this is equal to

$$\underbrace{\begin{bmatrix} r_{ss,n}(1) \\ r_{ss,n}(2) \\ \vdots \\ r_{ss,n}(P) \end{bmatrix}}_{\mathbf{r}_{ss,n}} = \underbrace{\begin{bmatrix} r_{ss,n}(0) & r_{ss,n}(1) & \cdots & r_{ss,n}(P-1) \\ r_{ss,n}(1) & r_{ss,n}(0) & \cdots & r_{ss,n}(P-2) \\ \vdots & \vdots & & \vdots \\ r_{ss,n}(P-1) & r_{ss,n}(P-2) & \cdots & r_{ss,n}(0) \end{bmatrix}}_{\mathbf{R}_{ss,n}} \underbrace{\begin{bmatrix} a_1 \\ a_2 \\ \vdots \\ a_P \end{bmatrix}}_{\mathbf{a}}. \tag{3.29}$$

Solving this equation for the optimal predictor coefficients a_i leads to

$$\mathbf{a} = \mathbf{R}_{ss,n}^{-1}\mathbf{r}_{ss,n}.$$
(3.30)

This equation is only solvable for $\mathbf{R}_{ss,n}$ being positive definite and therefore invertible. Since the matrix $\mathbf{R}_{ss,n}$ is symmetric and has *Toeplitz* structure, there exist very efficient algorithms to solve this set of equations [Rabiner 93]. The most popular one, the so-called *Levinson–Durbin recursion*, is presented in Appendix A.

3.1.2 Covariance Method

A second method is to use an unweighted speech signal that is not restricted to the segment chosen in (3.22) and instead restrict the computation of the mean-squared error to the range $m \in \{0, \ldots, N-1\}$. So we can write

$$e_{n,\sigma}(m) = \begin{cases} e_\sigma(n+m), & \text{for } m \in \{0, \ldots, N-1\}, \\ 0, & \text{otherwise,} \end{cases}$$
(3.31)

leading to the computation of the mean-squared error

$$E_n = \sum_{m=0}^{N-1} e_{n,\sigma}^2(m).$$
(3.32)

Analogous to the above considerations incorporating (3.19) we get

$$\phi_{ss,n}(i, \ell) = \sum_{m=0}^{N-1} s_n(m-i)s_n(m-\ell)$$
(3.33)

$$= \sum_{m=-i}^{N-1-i} s_n(m)s_n(m+i-\ell).$$
(3.34)

This means that speech samples outside the error computation interval are needed ($m \in \{-p, \ldots, -1\}$). If we use this modified definition of the covariance function $\phi_{ss,n}(i, \ell)$ we can express (3.20) in matrix-vector notation as

$$\begin{bmatrix} \phi_{ss,n}(1,0) \\ \phi_{ss,n}(2,0) \\ \vdots \\ \phi_{ss,n}(P,0) \end{bmatrix} = \begin{bmatrix} \phi_{ss,n}(1,1) & \phi_{ss,n}(1,2) & \cdots & \phi_{ss,n}(1,P) \\ \phi_{ss,n}(2,1) & \phi_{ss,n}(2,2) & \cdots & \phi_{ss,n}(2,P) \\ \vdots & \vdots & & \vdots \\ \phi_{ss,n}(P,1) & \phi_{ss,n}(P,2) & \cdots & \phi_{ss,n}(P,P) \end{bmatrix} \begin{bmatrix} a_1 \\ a_2 \\ \vdots \\ a_P \end{bmatrix}.$$
(3.35)

The matrix within (3.35) is still symmetric ($\phi_{ss,n}(i, \ell) = \phi_{ss,n}(\ell, i)$) but not *Toeplitz* any more. For the solution of this problem also very efficient algorithms like the so-called *Cholesky decomposition* exist [Rabiner 93]. Since this book respectively the algorithms presented within comprise only the autocorrelation method, the interested reader is referred to the literature [Markel 76]. Many improvements to the linear predictive analysis concerning the relation between formants and fundamental frequency have been investigated. Some approaches can be found in [El-Jaroudi 91, Murthi 00, Kabal 03].

3.2 Parametric Representations of the Spectral Envelope

In this section we will introduce several other well known representations of the vocal tract and the spectral envelope. Most of them depend on the above introduced linear predictive analysis or can be applied using the linear predictive coefficients as is the case with the linear predictive cepstral coefficients presented in Sect. 3.2.3. Even the tube model, presented below, yielding a completely different derivation motivation, has it's analogy in the linear predictive analysis when looking at the reflection coefficient. Also perceptually motivated representations of the spectral envelope will be introduced by presenting the mel frequency cepstral coefficients in Sect. 3.2.4. The section is intended to give a global overview on the possible representations of the filter part of the source-filter model which will help us finding the adequate parameter set for the bandwidth extension approaches in Chap. 5.

3.2.1 Tube Model

One possibility of modeling the vocal tract comprises the use of a tube [Vary 98]. This is one of the first attempts to describe a model for the speech production process and is briefly described here for historical reasons. The most general approach involves the use of a plain wave that propagates through a lossless cylindrical tube with a variable cross-sectional area $A(x)$

$$\frac{\partial^2 \Phi}{\partial x^2} + \frac{1}{A}\frac{\mathrm{d}A}{\mathrm{d}x}\frac{\partial \Phi}{\partial x} = \frac{1}{c^2}\frac{\partial^2 \Phi}{\partial t^2}, \tag{3.36}$$

which is known as *Webster's horn equation*, where c represents the sound-propagation velocity, x the longitudinal axis along the tube, t the time, and $\Phi(x,t)$ the so-called *velocity potential* which is defined by

$$p = \varrho_0 \frac{\partial \Phi}{\partial t} \tag{3.37}$$

and

$$v = -\frac{\partial \Phi}{\partial x}, \tag{3.38}$$

where ϱ_0 denotes the density of air in the tube, p the pressure, and v the sound particle velocity. As a rule, closed-form solutions for this problem are not available.

If we assume a model corresponding to Fig. 3.3 we can reformulate (3.36) for each tube segment with a constant cross-sectional area A_i to

$$\frac{\partial^2 \Phi_i}{\partial x^2} = \frac{1}{c^2}\frac{\partial^2 \Phi_i}{\partial t^2}. \tag{3.39}$$

For this problem, general solutions of the form

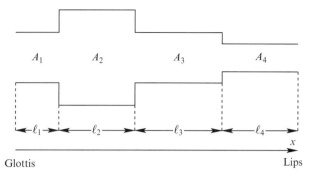

Glottis   Lips

Fig. 3.3. Model of the vocal tract consisting of a concatenation of several lossless cylindrical tubes of length ℓ_i and cross-sectional area A_i

$$\Phi_i(x,t) = \Phi_i^+ \left(t - \frac{x}{c} \right) + \Phi_i^- \left(t + \frac{x}{c} \right) \tag{3.40}$$

exist. Here Φ_i^+ denotes the wave that propagates in positive x-direction (forward) of segment i, whereas Φ_i^- denotes the wave that propagates in negative x-direction (backward) within the tube segment i, where x is measured from the glottal end of each tube ($0 \leq x \leq \ell_i$). Using (3.38) we can formulate the sound particle velocity as

$$v_i(x,t) = \frac{1}{c} \left[\Phi_i^+ \left(t - \frac{x}{c} \right) - \Phi_i^- \left(t + \frac{x}{c} \right) \right]. \tag{3.41}$$

In the case of one-dimensional airflow in the vocal tract, it is more convenient to look at the velocity of a volume of air u_i than its particle velocity v_i

$$u_i = A_i v_i. \tag{3.42}$$

Therefore we can further formulate

$$\begin{aligned}
u_i(x,t) &= \frac{A_i}{c} \left[\Phi_i^+ \left(t - \frac{x}{c} \right) - \Phi_i^- \left(t + \frac{x}{c} \right) \right] \\
&= u_i^+ \left(t - \frac{x}{c} \right) - u_i^- \left(t + \frac{x}{c} \right).
\end{aligned} \tag{3.43}$$

By defining the acoustic impedance

$$Z_i = \frac{\varrho_0 c}{A_i}, \tag{3.44}$$

we can write the solution for the pressure as

$$\begin{aligned}
p_i(x,t) &= \varrho_0 \left[\Phi_i^+ \left(t - \frac{x}{c} \right) + \Phi_i^- \left(t + \frac{x}{c} \right) \right] \\
&= p_i^+ \left(t - \frac{x}{c} \right) + p_i^- \left(t + \frac{x}{c} \right) \\
&= Z_i \left[u_i^+ \left(t - \frac{x}{c} \right) + u_i^- \left(t + \frac{x}{c} \right) \right].
\end{aligned} \tag{3.45} \tag{3.46}$$

Equation (3.46) is derived by inserting (3.42) into (3.38). Then (3.37) and (3.38) are solved for $\frac{\partial \Phi}{\partial t}$ and $\frac{\partial \Phi}{\partial x}$, respectively and the solutions are inserted into (3.39). Into this expression we can now insert (3.43) and (3.45). We can now eliminate the derivatives with respect to t by applying the relationship

$$\frac{\partial f(t \pm \frac{x}{c})}{\partial t} = \pm c \frac{\partial f(t \pm \frac{x}{c})}{\partial x}, \tag{3.47}$$

which is true for any differentiable function f. Equation (3.46) can then simply be obtained by comparing the coefficients.

If we now examine the continuity conditions at the transition of one tube segment i to the subsequent one $i+1$, we can express the volume velocity for the first case as

$$u_i(\ell_i, t) = u_i^+(t - \tau_i) - u_i^-(t + \tau_i), \tag{3.48}$$

where $\tau_i = \frac{\ell_i}{c}$ denotes the time needed by the wave to propagate through the tube segment i with length ℓ_i. In the same manner we can express the volume velocity for the second case

$$u_{i+1}(0, t) = u_{i+1}^+(t) - u_{i+1}^-(t). \tag{3.49}$$

In a physical system the volume velocity and the pressure have to be continuous at the transition of tube segment i to tube segment $i+1$, resulting in

$$u_i(\ell_i, t) = u_{i+1}(0, t), \tag{3.50}$$
$$p_i(\ell_i, t) = p_{i+1}(0, t). \tag{3.51}$$

Inserting (3.43) and (3.46) into (3.50) and (3.51), respectively, leads to

$$u_i^+(t - \tau_i) - u_i^-(t + \tau_i) = u_{i+1}^+(t) - u_{i+1}^-(t), \tag{3.52}$$
$$u_i^+(t - \tau_i) + u_i^-(t + \tau_i) = \left[u_{i+1}^+(t) + u_{i+1}^-(t)\right] \frac{A_i}{A_{i+1}}. \tag{3.53}$$

If we now assume the case of $u_{i+1}^-(t) = 0$, then $-u_i^-(t + \tau_i)$ can be interpreted as the reflection of $u_i^+(t - \tau_i)$ at the transition of tube segment i to $i+1$. Setting $u_{i+1}^-(t) = 0$ in (3.52) and (3.53) leads, after combining the two equations, to

$$-u_i^-(t + \tau_i) = \frac{A_{i+1} - A_i}{A_{i+1} + A_i} u_i^+(t - \tau_i). \tag{3.54}$$

Thus we can define the reflection coefficient[2] as

$$r_i = \frac{A_{i+1} - A_i}{A_{i+1} + A_i}. \tag{3.55}$$

---

[2] Note that in contrast to the derivation of the Levinson–Durbin algorithm in Appendix A the numbering of the elements A_i appears here in the opposite direction (from the glottis to the mouth).

Since the cross sectional areas possess always positive values the reflection coefficient is limited to

$$-1 \leq r_i \leq 1. \tag{3.56}$$

The boundary values arise for the case of an opened ($A_{i+1} = \infty$, $Z_{i+1} = 0$, $r_i = 1$) and a closed end ($A_{i+1} = 0$, $Z_{i+1} = \infty$, $r_i = -1$), respectively. The reflection coefficient becomes zero for $A_i = A_{i+1}$ which means $Z_i = Z_{i+1}$. We will encounter this reflection coefficient once again when discussing the Levinson–Durbin recursion for solving the normal equations (3.30) in Appendix A using (A.18).

For the use within a discrete model this tube model gets especially applicable when setting the individual tube lengths ℓ_i to equal lengths. Thereby we obtain a sampled version of the cross sectional area function of the vocal tract. This leads to the well known *Kelly–Lochbaum structure*. The tube model is one of the first approaches to model the vocal tract and therewith being able to synthesize speech. However this historical approach involves unjustifiable use of resources in terms of computing power and is therefore presented only to complement the view and to show the analogy within the linear predictive analysis.

3.2.2 AR-Coefficients

A term that is used rather frequently in speech coding (see [Atal 82]) or speech analysis is the so-called *AR-model* (auto-regressive model) or the *AR-coefficients*, respectively. This concept is based on modeling the mouth and nasal cavities using a discrete, time variant filter. In Sect. 2.2.2 we already introduced the approach of modeling the influence of the vocal tract on the excitation signal by a discrete, time variant filter. A general formulation for the transfer function of such a filter is

$$H(z) = \frac{B(z)}{A(z)} = \frac{\sum\limits_{i=0}^{P} b_i z^{-i}}{\sum\limits_{i=0}^{P} a_i z^{-i}}. \tag{3.57}$$

This represents the so-called ARMA-model (auto-regressive moving average). As special cases the MA-model would be characterized by $a_i = 0$ for $i \in \{1, \ldots, P\}$, and possesses only zeros in the z-plane, whereas the AR-model would be characterized by $b_i = 0$ for $i \in \{1, \ldots, P\}$, and is therefore a straight recursive filter and possesses only poles in the z-plane.

When analyzing (3.5) we observe that the approach of modeling the vocal tract transfer function by using an AR-model has already been addressed under the term *linear predictive analysis* in Sect. 3.1. Since the approach presented in Sect. 3.1 is very intuitive and has already been described in detail we will equivalently use the terms *LP-coefficients* and *AR-coefficients* for the rest of the book.

3.2.3 Cepstral Coefficients

The speech production according to the source-filter model produces a speech signal that results from the convolution of an excitation signal with the impulse response of the vocal tract. This convolution makes it difficult to extract or influence only one of these source signals or parameters, respectively [O'Shaughnessy 00]. The application of the logarithm carries multiplications over to summations. This motivates the definition of the *complex cepstrum* as the inverse discrete time Fourier transform of the natural logarithm of the speech spectrum

$$c_i = \mathcal{F}^{-1}\left\{\ln\left(\mathcal{F}\left\{s(n)\right\}\right)\right\} \tag{3.58}$$

$$c_i = \mathcal{F}^{-1}\left\{\ln\left(S\left(e^{j\Omega}\right)\right)\right\} \tag{3.59}$$

$$\mathcal{F}\left\{c_i\right\} = \ln\left(S\left(e^{j\Omega}\right)\right), \tag{3.60}$$

where $\mathcal{F}$ denotes the discrete time Fourier transform (DTFT)

$$S\left(e^{j\Omega}\right) = \mathcal{F}\left\{s(n)\right\} = \sum_{n=-\infty}^{\infty} s(n)e^{-jn\Omega}, \tag{3.61}$$

and $\mathcal{F}^{-1}$ the inverse discrete time Fourier transform (IDTFT)

$$s(n) = \mathcal{F}^{-1}\left\{S\left(e^{j\Omega}\right)\right\} = \frac{1}{2\pi}\int_{-\pi}^{\pi} S\left(e^{j\Omega}\right)e^{jn\Omega}d\Omega, \tag{3.62}$$

respectively. Equation (3.59) is called the complex cepstrum since the speech spectrum is defined as a complex signal. By applying the ln on the complex spectrum we do have to define a complex ln

$$\ln(z) = \ln|z| + j\arg\{z\}, \tag{3.63}$$

where $\arg\{z\}$ denotes the angle of z within the complex plain. The resulting cepstral coefficients are real due to the special symmetry properties of $S\left(e^{j\Omega}\right)$. Equation (3.59) leads to the formulation

$$\sum_{i=-\infty}^{\infty} c_i e^{-j\Omega i} = \ln\left(S(e^{j\Omega})\right). \tag{3.64}$$

Alternatively to (3.59) the cepstral coefficients used in this book are so-called *linear predictive cepstral coefficients*. As indicated by the name these coefficients are computed on the basis of linear predictive coefficients. By exchanging the speech spectrum $S(e^{j\Omega})$ by its representation using an all-pole model we obtain

$$\ln\left(S\left(e^{j\Omega}\right)\right) = \sum_{i=-\infty}^{\infty} c_i z^{-i}\Bigg|_{z=e^{j\Omega}} \tag{3.65}$$

$$\approx \ln\left[\frac{\sigma}{A(z)}\right]\Bigg|_{z=e^{j\Omega}} \tag{3.66}$$

$$= \ln\sigma - \ln\left(A(z)\right)|_{z=e^{j\Omega}}. \tag{3.67}$$

Let us now have a closer look at the last term $\ln\left(A(z)\right)$

$$\ln\left(A(z)\right) = \ln\left[\sum_{k=0}^{P} a_k z^{-k}\right] \tag{3.68}$$

by representing this expression as a product of roots of the polynomial with modified coefficients u_k we can write

$$\ln\left[\sum_{k=0}^{P} a_k z^{-k}\right] = \ln\left[\prod_{k=0}^{P}(1 - u_k z^{-1})\right] \tag{3.69}$$

$$= \sum_{k=0}^{P} \ln\left[1 - u_k z^{-1}\right]. \tag{3.70}$$

By exploiting the following series expansion [Bronstein 85]:

$$\ln\left[1 - \alpha z^{-1}\right] = -\sum_{n=1}^{\infty} \frac{\alpha^n}{n} z^{-n}, \qquad |z| > |\alpha|, \tag{3.71}$$

that holds for factors that converge within the unit circle, which is the case here since $A(z^{-1})$ is analytic inside the unit circle [Oppenheim 89], we can further rewrite (3.70)

$$\ln\left(A(z)\right) = \ln\left[\sum_{k=0}^{P} a_k z^{-k}\right] = -\sum_{k=0}^{P}\sum_{n=1}^{\infty} \frac{u_k^n}{n} z^{-n} \tag{3.72}$$

$$= -\sum_{n=1}^{\infty}\underbrace{\sum_{k=0}^{P} \frac{u_k^n}{n} z^{-n}}_{c_i}. \tag{3.73}$$

If we now compare (3.73) with (3.65) we observe that the two sums do not have equal limits. This means that the c_i are equal to zero for $i < 0$. For $i > 0$ we can set the c_i equal to the second sum in (3.73). For $i = 0$ we have $c_0 = \ln\sigma$ from (3.67). So we can state concluding the above observations

$$c_i \stackrel{!}{=} \begin{cases} \sum_{k=0}^{P} \frac{u_k^i}{i}, & \text{for } i > 0, \\[2mm] \ln\sigma, & \text{for } i = 0, \\[2mm] 0, & \text{for } i < 0. \end{cases} \tag{3.74}$$

This leads to the assertion

$$\ln\left(A(z)\right) = -\sum_{i=1}^{\infty} c_i z^{-i}, \tag{3.75}$$

which means

$$\ln\left[\frac{1}{A(z)}\right] = \sum_{i=1}^{\infty} c_i z^{-i}, \tag{3.76}$$

where we can finally add the generic model gain from (3.67) resulting in

$$\ln\left[\frac{\sigma}{A(z)}\right] = \ln\sigma + \sum_{i=1}^{\infty} c_i z^{-i}. \tag{3.77}$$

If the P predictor coefficients are known we can derive a simple recursive computation of e.g. $m = \frac{3}{2}P$ linear predictive cepstral coefficients by differentiating both sides of (3.77) with respect to z and equating the coefficients of like powers of z:

$$\ln\left[\frac{\sigma}{A(z)}\right] = \ln\sigma + \sum_{k=1}^{\infty} c_k z^{-k}$$

$$-\frac{d}{dz}\ln\left[1 - \sum_{i=1}^{P} a_i z^{-i}\right] = \frac{d}{dz}\sum_{k=1}^{\infty} c_k z^{-k} \tag{3.78}$$

$$\sum_{i=1}^{P} i a_i z^{-i-1}\left[-1 + \sum_{i=1}^{P} a_i z^{-i}\right]^{-1} = -\sum_{k=1}^{\infty} k c_k z^{-k-1} \tag{3.79}$$

$$\sum_{i=1}^{P} i a_i z^{-i-1} = \sum_{k=1}^{\infty} k c_k z^{-k-1} - \sum_{k=1}^{\infty}\sum_{i=1}^{P} k c_k a_i z^{-k-i-1} \tag{3.80}$$

If we now consider the equation above for equal powers of z, we find that starting from left, the first two terms only contribute a single term each up to z^{-P-1}. We will label the order with i. The last term in contrast produces an amount of terms that depend on i

$$i a_i = i c_i - \sum_{k=1}^{i-1} k c_k a_{i-k}, \qquad i \in \{1,\ldots,P\}, \tag{3.81}$$

$$c_i = a_i + \frac{1}{i}\sum_{k=1}^{i-1} k c_k a_{i-k}, \qquad i \in \{1,\ldots,P\}. \tag{3.82}$$

For $k > P$ in (3.80) the first term on the right side still needs to be considered whereas the term on the left side of (3.80) does not contribute any more to the c_i in (3.82) and can therefore be omitted meaning for $i > P$ in (3.82)

$$c_i = \frac{1}{i} \sum_{k=1}^{i-1} k c_k a_{i-k}, \quad i > P. \tag{3.83}$$

Summarizing (3.82) and (3.83) we can formulate a recursive computation of cepstral coefficients from linear predictive coefficients

$$c_i = \begin{cases} a_i + \frac{1}{i} \sum_{k=1}^{i-1} k c_k a_{i-k}, & i \in \{1, \dots, P\}, \\ \frac{1}{i} \sum_{k=1}^{i-1} k c_k a_{i-k}, & i > P, \\ \ln \sigma^2, & i = 0, \\ 0, & i < 0. \end{cases} \tag{3.84}$$

This leads to $c_0 = 0$ for a normalized envelope spectrum with $\sigma = 1$. There exist also non-recursive relations between cepstral coefficients and predictor coefficients [Schroeder 81].

Cepstral coefficients are widely used in the field of speech recognition. They are known to be more robust against distortions than the above presented LP-coefficients. Furthermore, concerning bandwidth extension, a well known distance measure depending on the mean squared cepstral distance exists, as will be presented in Sect. 3.4.3. This simple and effective distance measure as well as the robustness properties make cepstral coefficients an adequate choice for the application within neural network approaches since the learning algorithm fits with the cepstral distance measure (see Sect. 5.4). But also within codebook and linear mapping approaches (see Sect. 5.3 and Sect. 5.5) cepstral coefficients have been employed successfully. In Appendix E the distributions for the cepstral coefficients used in Sect. 5.3.2 are shown.

3.2.4 MFCCs

Another set of coefficients that is nowadays the most popular choice in speech recognition applications are the so-called *mel frequency cepstral coefficients* (MFCC) [Seltzer 05a, Valin 00]. The MFCCs are computed analog to the cepstral coefficients. The only difference lies in the non-linear frequency scale. This *warped* frequency scale is based on the so-called mel scale which is motivated by psycho acoustics. The mel scale originates from experiments with subjects that have been asked to adjust a sine tone to the double frequency as the 1 kHz reference sine tone they were listening to, the half, ten times etc. The resulting relation between the real frequency and the one that has been set by the subjects is depicted in Fig. 3.4 [Rabiner 93].

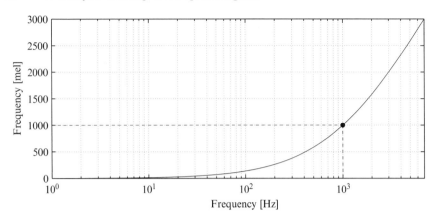

Fig. 3.4. The mel scale

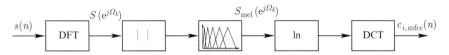

Fig. 3.5. Block diagram of MFCC computation

The mel-spectrum and afterwards the MFCCs used in this book have been computed following the block diagram in Fig. 3.5 which represents the standard procedure described in the literature [Deller Jr. 00, Rabiner 93]. First the speech segment $s(n)$ is transformed into the frequency domain by a discrete Fourier transform (DFT)

$$S\left(e^{j\Omega_k}\right) = \sum_{n=0}^{N-1} s(n)e^{-jn\Omega_k}, \tag{3.85}$$

with

$$\Omega_k = \frac{2\pi k}{N}. \tag{3.86}$$

Then the absolute value of the spectrum is filtered by a triangular shaped mel-filterbank which can be seen in Fig. 3.6 for an example with $N_{\mathrm{mel}} = 12$ mel-filters. After the mel-filterbank the spectral components $S_{\mathrm{mel}}\left(e^{j\Omega_k}\right)$ are reduced to the order of the filterbank. In this book $N_{\mathrm{mel}} = 28$ mel-filters have been used. Then the ln of the mel-spectrum is taken and the following IDFT is replaced due to the symmetric and real characteristic of the mel-spectrum by a discrete cosine transform (DCT)

$$c_{i,\mathrm{mfcc}} = w_i \cdot \sum_{k=0}^{N_{\mathrm{mel}}-1} \ln\left(S_{\mathrm{mel}}\left(e^{j\Omega_k}\right)\right) \cdot \cos\left(\frac{i(k+0.5)\pi}{N_{\mathrm{mel}}}\right), \tag{3.87}$$

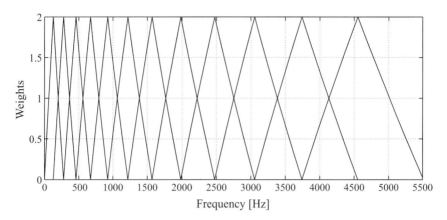

Fig. 3.6. Mel-filterbank

for $i \in \{0, \ldots, N_{\mathrm{mfcc}} - 1\}$, where

$$w_i = \begin{cases} \sqrt{2} \,, & \text{for } i = 0 \\ 2 \,, & \text{otherwise.} \end{cases} \tag{3.88}$$

The order of the DCT (N_{mfcc}) then determines the amount of MFCCs.

In [Enbom 99, Raza 03] MFCCs are used for a bandwidth extension application. However MFCCs have not justified the extra amount of computational complexity needed by the additional DFT, filterbank and IDFT, compared to the use of LPC coefficients, by improved quality as will be presented in Sect. 6.2.1. For speech recognition tasks MFCCs have proven to outperform LPC coefficients and cepstral coefficients which does not conflict the assertion that MFCCs are not the optimal choice for bandwidth extension tasks since for speech recognition the characteristics of a given voice are not of interest for identifying an utterance in contrast to bandwidth extension schemes where the characteristics of a given voice should be retained. Additionally speech recognition operates off line and is therefore not as critical as bandwidth extension schemes concerning computational complexity.

3.2.5 Line Spectral Frequencies

Another set of coefficients that is derived from LP-coefficients are the so-called *line spectral frequencies* or equivalent *line spectrum pairs*. These coefficients are very popular in speech coding as well as in bandwidth extension applications (see [Chan 96, Chan 97, Chen 04, Qian 03, Yao 06]) since they have a couple of interesting properties concerning quantizing and stability issues as we will see later on. Starting point for the considerations is the z-domain representation of the inverse filter of order N,

$$A(z) = 1 - \sum_{i=1}^{N} a_i z^{-i}. \tag{3.89}$$

This polynomial can be split into two order $N+1$ polynomials

$$P(z) = A(z) + A(z^{-1})z^{-(N+1)}, \tag{3.90}$$
$$Q(z) = A(z) - A(z^{-1})z^{-(N+1)}, \tag{3.91}$$

where $P(z)$ is a mirror polynomial and $Q(z)$ an anti-mirror polynomial. The primal expression is then obtained by

$$A(z) = \frac{P(z) + Q(z)}{2}. \tag{3.92}$$

The mirror property is characterized by

$$P(z) = z^{-(N+1)}P(z^{-1}), \tag{3.93}$$
$$Q(z) = z^{-(N+1)}Q(z^{-1}). \tag{3.94}$$

As shown in [Schüßler 94] $A(z)$ is guaranteed to be a minimum phase filter (and therefore the synthesis filter $H(z) = 1/A(z)$ is guaranteed to be a stable filter) if the zeros of $P(\lambda_{i,P}) = 0$ and $Q(\lambda_{i,Q}) = 0$ lie on the unit circle and if they are increasing monotonously and alternating as shown in Fig. 3.7 for an all-pole filter $H(z)$ of order $N = 8$. In fact $P(z)$ has a real zero at $z = -1$ and $Q(z)$ at $z = 1$. Since $H(z)$ is an all-pole filter, the above mentioned constraints are fulfilled. Now it is straight forward to design a stable filter if we assume that all zeros lie on the unit circle. So we have to quantize only the angles α_i, which represent the LPC coefficients, of the zeros λ of $P(z)$ and $Q(z)$ in an alternating and monotonously increasing manner

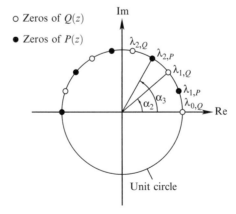

Fig. 3.7. Example for the zeros λ of the LSF polynomials $P(z)$ and $Q(z)$ for $N = 8$

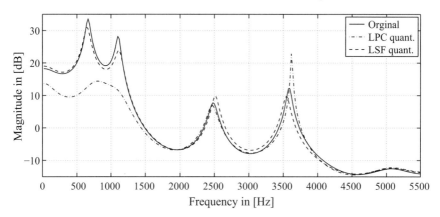

Fig. 3.8. An original envelope and two versions quantized with 6 bit

$$\alpha_i = \arctan \frac{\Im\{\lambda_{i,P,Q}\}}{\Re\{\lambda_{i,P,Q}\}} \text{ , for } i = 0, \ldots, N-1, \tag{3.95}$$

with N being the order of the all-pole filter $\frac{1}{A(z)}$. In Fig. 3.8 two versions of an original spectral envelope quantized with 6 bit are depicted. The representation using LPC coefficients as well as LSF coefficients have been used. It is obvious that the LSF coefficients are more robust against quantization as the LPC coefficients [Soong 93]. Another aspect that is more important in bandwidth extension approaches is the more robust characteristic of LSFs compared to other parametric representations of the spectral envelope when building a codebook entry by quantizing an amount of feature vectors to a cluster and succeeding derivation of the centroid by summing the feature vectors up and calculating the mean vector. Here LSFs also show a more robust behavior. However the benefit concerning robustness gained by the usage of LSFs in comparison to the increased computational complexity by searching the zeros of the polynomials does not justify the usage of LSFs (see Sect. 6.2.1).

3.3 Scalar Speech Features

After the description of the parametric representation of the spectral envelope we will now have a closer look on scalar speech features. These features are most often independent of the spectral envelope. Some of them indicate if the current speech frame contains a voiced or an unvoiced utterance. Some others indicate the gender of the speaker. This can be exploited concerning bandwidth extension algorithms by using different models or extension schemes based on a voiced unvoiced or gender detection and can therefore increase the robustness against malclassifications [Raza 02a, Raza 02b]. We will present the most common features and discuss briefly their physical interpretation.

3.3.1 Zero Crossing Rate

The short-term zero crossing rate for a block of N speech samples starting at sample m is defined as

$$Z_s(m) = \frac{1}{2(N-1)} \sum_{n=m+1}^{m+N-1} |\operatorname{sgn}\{s(n)\} - \operatorname{sgn}\{s(n-1)\}|, \qquad (3.96)$$

where

$$\operatorname{sgn}\{s(n)\} = \begin{cases} +1, & s(n) \geq 0 \\ -1, & s(n) < 0 \end{cases}. \qquad (3.97)$$

The zero crossing rate is an indicator for the presence of voiced or unvoiced speech, respectively. For unvoiced speech the zero crossing rate shows in general higher values than for unvoiced speech. In Fig. 3.9 an example for a voiced and an unvoiced block of $N = 256$ samples is given. However the analysis of the zero crossing rate should be combined with a speech activity detector since in the presence of noise it is unlikely to be able to differentiate between a fricative and noise for example.

3.3.2 Gradient Index

The so-called *gradient index* is defined similarly to the zero crossing rate. The main differences are the dependence on the magnitude and the use of the so-called *change of direction*. The gradient index is defined as

$$x_{s,\mathrm{gi}}(m) = \frac{\sum\limits_{n=m+1}^{m+N-1} \Psi(n)\,|s(n) - s(n-1)|}{\sqrt{\frac{1}{N} \sum\limits_{n=m}^{m+N-1} s^2(n)}}, \qquad (3.98)$$

where

$$\Psi(n) = \frac{1}{2}\,|\psi(n) - \psi(n-1)|. \qquad (3.99)$$

Here $\psi(n)$ denotes the sign of the gradient $\psi(n) = \operatorname{sgn}\{s(n) - s(n-1)\}$ [Jax 04a]. As well as the zero crossing rate, the gradient index shows higher values for unvoiced speech segments and lower values for voiced ones (see Fig. 3.10).

3.3.3 Fundamental Frequency

Another interesting scalar speech feature is the fundamental frequency (pitch frequency or simply pitch). As already mentioned in Chap. 2 it is correlated with the gender of the speaker. One possible approach for extracting the pitch frequency of a speech signal is called autocorrelation method. Since voiced

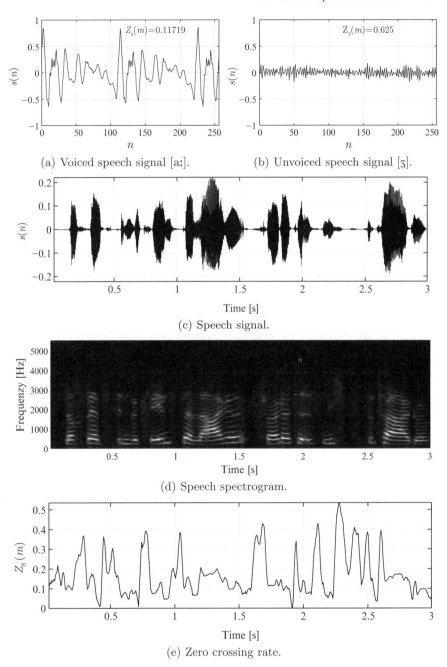

(a) Voiced speech signal [aː].

(b) Unvoiced speech signal [ʒ].

(c) Speech signal.

(d) Speech spectrogram.

(e) Zero crossing rate.

Fig. 3.9. (**a**) and (**b**) Show one block of a voiced and an unvoiced sequence of a speech signal and their respective zero crossing rates; (**c**) and (**d**) show a speech signal and its spectrogram, respectively. In (**e**) the resulting progression of the zero crossing rate is depicted

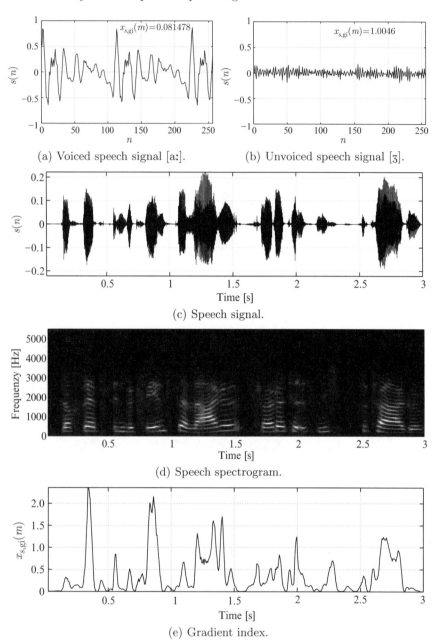

(a) Voiced speech signal [aː].

(b) Unvoiced speech signal [ʒ].

(c) Speech signal.

(d) Speech spectrogram.

(e) Gradient index.

Fig. 3.10. (a) and (b) Show one block of a voiced and an unvoiced sequence of a speech signal and their respective gradient index; (c) and (d) show a speech signal and its spectrogram, respectively. In (e) the resulting progression of the gradient index is depicted

signals comprise periodic signals the periodicity should become apparent as a local maximum within the autocorrelation of the signal. One major problem using this method is that also multiples of this periodicity become apparent and have sometimes even a higher magnitude than the peak at the real periodicity. Therefore it is reasonable to look for the first local maximum starting from longer periodicities.

In the following the method that has been derived and is used in this book is described. In a first step a modified short-term autocorrelation is computed

$$r_{ss,\mathrm{mod}}(i) = \frac{1}{\sqrt{w(i)}} \sum_{k=0}^{N-1-i} s_n(k)s_n(k+i) \qquad i \in \{0, \dots, N-1\}. \qquad (3.100)$$

This once again premises the definition of a short segment of speech

$$s_n(m) = s(n+m), \quad m \in \{0, \dots, N-1\}. \qquad (3.101)$$

The modification lies in the multiplication with a term that reduces the double windowing influence to a single one. The reason for this modification is that for block processing e.g., a Hann window is used. So the autocorrelation function is a product of a Hann window and a shifted version of this window weighting the underlying autocorrelation. This would increasingly attenuate the magnitude of local maxima with longer periodicity and therefore reduce the probability of detecting lower pitch frequencies. To avoid this, a factor has been introduced weighting the autocorrelation function.

Let us start beginning with a Hann window as used for windowing the speech segments

$$h_{\mathrm{hann}}(i) = \frac{1}{2}\left(1 - \cos\left(\frac{2\pi i}{N}\right)\right) \qquad i \in \{0, \dots, N-1\}. \qquad (3.102)$$

Then we can define the autocorrelation of this window as

$$w_{hh}(i) = \frac{1}{\sum\limits_{k=0}^{N-1} h_{\mathrm{hann}}^2(k)} \sum_{k=0}^{N-1-i} h_{\mathrm{hann}}(k)h_{\mathrm{hann}}(k+i) \qquad i \in \{0, \dots, N-1\}.$$

$$(3.103)$$

The function is limited to a minimum of ϵ so the autocorrelation function is not too much amplified for bigger time lags

$$w(i) = \max\{w_{hh}(i), \epsilon\} \qquad i \in \{0, \dots, N-1\}. \qquad (3.104)$$

The resulting $w(n)$ for $\epsilon = 0.2$ is shown in Fig. 3.11. This modification is intended to prevent for example the well known problem of pitch doubling in current pitch determination algorithms.

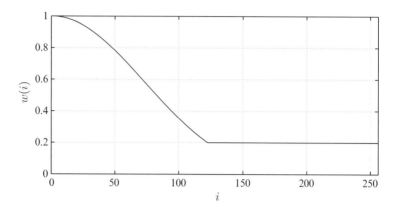

Fig. 3.11. Modified autocorrelation $w(i)$ of a Hann window for $\epsilon = 0.2$

Now for extracting the pitch we are looking for the local maximum within the range where we expect the pitch.

$$ACI = i_{\max} = \arg \max_{i=i_{\text{low}}}^{i_{\text{high}}} \{r_{ss,\text{mod}}(i)\} \qquad (3.105)$$

Besides the index where the maximum within the short-time correlation function can be found, the value of this maximum can be used to either decide whether the utterance is voiced or unvoiced or how reliable the pitch detection was

$$ACM = \frac{r_{ss,\text{mod}}(i_{\max})}{r_{ss,\text{mod}}(0)\sqrt{w(i_{\max})}}. \qquad (3.106)$$

In Fig. 3.12 an example for the here described pitch estimation algorithm is presented. Note that the estimated pitch value has been set to zero for $ACM < 0.55$. This means ACM has been used as a measure for the reliability and as a voiced/unvoiced classification at the same time. Part of the difference that may be observable in Fig. 3.12 between the estimated and the real pitch is due to the order of the DFT. For the speech spectrogram in Fig. 3.12 a DFT with a much higher resolution has been used as for the pitch estimation is achievable with the used block length. An example for the distributions of the values for ACI and ACM can be found in Appendix E.

Alternatively the pitch estimation can be performed in the frequency domain. Here we compare a set of products consisting of equidistant frequency bins of the spectrum of the excitation signal. For each set the distance is increased by a few Hz. Then we are looking for the maximum of the products[3]

---

[3] For keeping the equation in a form that remains readable we have dropped the notation used up to now and we will write $E(k)$ instead of $E\left(e^{j\Omega_k}\right)$.

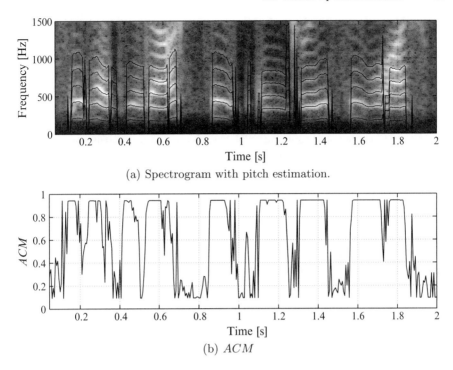

(a) Spectrogram with pitch estimation.

(b) *ACM*

Fig. 3.12. (a) Shows the spectrogram of a speech signal with the estimated pitch and the first four harmonics; (b) shows the magnitude of the normalized maximum within the autocorrelation (*ACM*)

$$\hat{k}_{\text{pitch}} = \arg\max_{k} \left\{ \frac{\prod_{\ell=0}^{M-1} \left| E\left(\left\lceil \frac{k_{300\text{Hz}}}{k} \right\rceil k + \ell k \right) \right|}{\prod_{\ell=0}^{M-1} \left| E\left(\left\lceil \frac{k_{300\text{Hz}}}{k} \right\rceil k + \ell k + \left[\frac{k}{2} \right] \right) \right|} \right\}, \tag{3.107}$$

where $k > 1$ is the frequency bin index where we are searching for pitch (between 50 Hz and 350 Hz) and $k_{300\text{Hz}}$ the bin that corresponds to the lower cut-off frequency of the telephone bandpass (300 Hz). The amount of factors is M. The resulting frequency index should neither exceed half of the DFT order, nor the upper cut off frequency of the telephone bandpass. The operators $\lceil x \rceil$ and $[x]$ denote the next natural number $i \in \mathbb{N}$ with $i \geq x$ and the nearest natural number to x, respectively. Using this expression we are able to use quasi-continuous values where we are looking for pitch. However the product is only evaluated at discrete frequency bins delivered by the DFT. This method is called *harmonic product spectrum*. The approach presented in (3.107) is a modified version which evaluates the ratio of the product where we are looking for pitch to the product of the frequency bins that are right between those we are looking for pitch. By the maximization of this ratio we ensure that we do not obtain pitch doubling, which is a problem in current algorithms, since we would have local maxima between the supposed ones, which decreases the ratio.

3.3.4 Kurtosis

The kurtosis originates from higher-order statistics and is computed using the fourth and second-order moments of the signal. The kurtosis measures the bulge of a surface for example and is defined as

$$x_{s,\mathrm{lk}}(m) = \frac{\mathcal{E}\left\{(s(m) - \bar{s}(m))^4\right\}}{\sigma_s^4(m)}, \tag{3.108}$$

where $\bar{s}(m)$ denotes the mean value of $s(m)$ and $\sigma_s(m)$ the standard deviation. An estimate for the *local* kurtosis [Jax 04a] under the assumption of a zero-mean process can be obtained using

$$x_{s,\mathrm{lk}}(m) = \frac{\frac{1}{N}\sum\limits_{n=m}^{m+N-1}(s(n))^4}{\left(\frac{1}{N}\sum\limits_{n=m}^{m+N-1}(s(n))^2\right)^2}. \tag{3.109}$$

The kurtosis is dimensionless and indicates the 'Gaussianity' of a random signal. Optimal Gaussianity would lead to $x_{s,\mathrm{lk}} = 3$, which would be the case for a Gaussian random variable. For most voiced speech segments $x_{s,\mathrm{lk}}$ would be less than 3. However at the onset of plosives and strong vowels $x_{s,\mathrm{lk}}$ reaches values that lie significantly above 3. Figure 3.13 shows the local kurtosis values for a voiced and an unvoiced segment of speech of length $N = 256$. As can be seen the value of the local kurtosis for the voiced speech segment lies over 3 which indicates a strong vowel. The value of the kurtosis for the unvoiced utterance however lies below 3. Experiments using the local kurtosis showed difficulties in a robust voiced/unvoiced classification and have therefore been abandoned within this work.

3.3.5 Spectral Centroid

The spectral centroid is a measure that indicates where most of the power of a speech segment is spectrally located. The spectral centroid is defined as

$$x_{s,\mathrm{sc}} = \frac{\sum\limits_{k=0}^{N/2} k \cdot \left|S\left(e^{j\Omega_k}\right)\right|}{\left(\frac{N}{2} + 1\right)\sum\limits_{k=0}^{N/2}\left|S\left(e^{j\Omega_k}\right)\right|}. \tag{3.110}$$

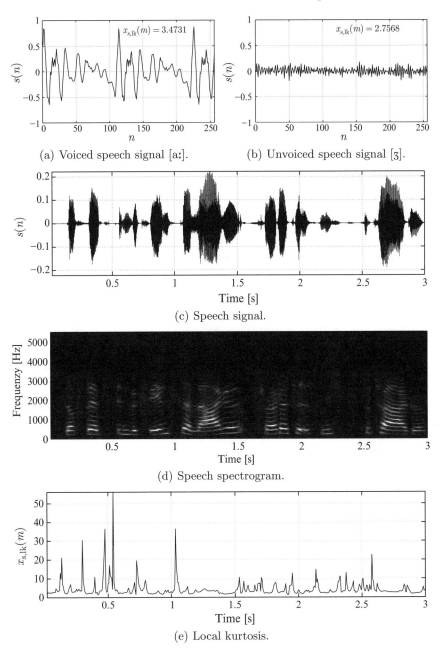

(a) Voiced speech signal [aː].

(b) Unvoiced speech signal [ʒ].

(c) Speech signal.

(d) Speech spectrogram.

(e) Local kurtosis.

Fig. 3.13. (a) and (b) Show one block of a voiced and an unvoiced sequence of a speech signal and their respective local kurtosis values; (c) and (d) show a speech signal and its spectrogram, respectively. In (e) the resulting progression of the local kurtosis is depicted

The spectral centroid lies between zero and one, depending mostly on the voicedness of the speech segment. For voiced segments values around 0.28 corresponding to 1,500 Hz are very common assuming a sampling rate of 11,025 Hz [Jax 04a]. Figure 3.14 shows an example for the behavior of the spectral centroid feature.

3.3.6 Energy Based Features

Energy based features are interesting as they allow to differentiate between speech activity and pauses. Another interesting observation is that also fricatives and vowels can be distinguished by their relative power within a speech segment to the long term mean power estimation. This long term mean power estimation has to be implemented adaptively. Vowels in general have more power than fricatives. Energy based features can not only be used in a scalar manner as the mean over the entire spectrum but also as a spectral vector. The significance of energy based features strongly depends on the SNR of the signal. Therefore it is reasonable to implement a noise estimator too.

Let us start with the short-term energy of a segment of speech with length N starting at time m

$$W(m) = \sum_{n=m}^{m+N-1} s^2(n). \tag{3.111}$$

The long-term average can then be estimated using a simple first order IIR filter

$$\bar{W}(m) = \alpha \bar{W}(m-N) + (1-\alpha)W(m). \tag{3.112}$$

A noise estimation can be implemented by observing the minimum within a frame

$$W_{\text{noise}}(m) = \left(\min_{n=m}^{m+N-1} [W(n), W_{\text{noise}}(m-N)] \right)(1+\epsilon). \tag{3.113}$$

Further this minimum is combined with the noise estimation of the previous frame. For being able to follow an increasing noise floor we add a multiplication by a factor $1 + \epsilon > 1$. An alternative approach that delivers better noise estimates at much higher computational complexity can be found in [Martin 01].

Normalized Frame Energy

By normalizing the short-term energy by the long-term estimate, we get independent of energy differences due to different speakers or different recording or transmission equipment as well as of different signal representations

$$W_{\text{NFE}}(m) = 10 \log_{10} \frac{W(m)}{\bar{W}(m)}. \tag{3.114}$$

Note that $W_{\text{NFE}}(m)$ already represents a dB value although it is labeled similar to the latter three energy estimations.

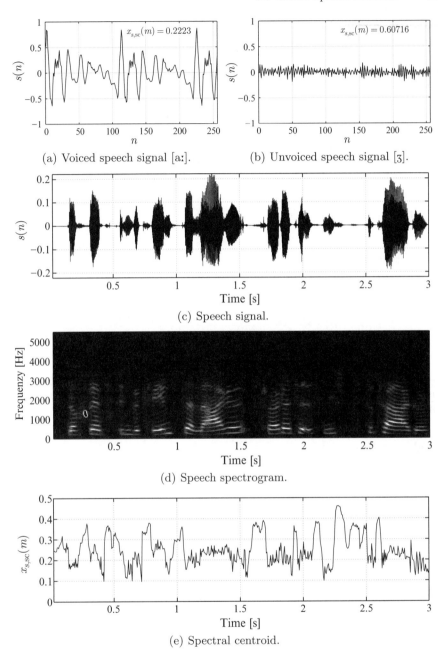

(a) Voiced speech signal [aː].

(b) Unvoiced speech signal [ʒ].

(c) Speech signal.

(d) Speech spectrogram.

(e) Spectral centroid.

Fig. 3.14. (**a**) and (**b**) Show one block of a voiced and an unvoiced sequence of a speech signal and their respective spectral centroid; (**c**) and (**d**) show a speech signal and its spectrogram, respectively. In (**e**) the resulting progression of the spectral centroid is depicted

Relative Frame Energy

The relative frame energy depends on the noise estimation and is favorably employed when we have to deal with background noise as it is the case when running the application in a car

$$W_{\mathrm{RFE}}(m) = 10 \log_{10} \frac{W(m)}{W_{\mathrm{noise}}(m)}. \tag{3.115}$$

In parallel to the above presented normalized frame energy $W_{\mathrm{RFE}}(m)$ represents a dB value and is also known as signal to noise ratio (SNR).

Normalized Relative Frame Energy

A mixture of the above presented energy based features combines as well the long-term energy estimate as well as the estimate for the noise energy resulting in a feature that is independent on short-term energy variations as well as of the noise level

$$r_{\mathrm{NRFE}}(m) = \frac{10 \log_{10} W(m) - 10 \log_{10} W_{\mathrm{noise}}(m)}{10 \log_{10} \bar{W}(m) - 10 \log_{10} W_{\mathrm{noise}}(m)}. \tag{3.116}$$

Figure 3.15 shows the progression of the energy based features that have been introduced up to now applied on an exemplary speech signal. In the example presented in Fig. 3.15 the different energy based speech features do not differ much since the example contained virtually no noise.

Highpass Energy to Lowpass Energy Ratio

Another interesting energy based feature is the highpass energy to lowpass energy ratio. This feature is able to indicate whether the current utterance has a rather voiced or unvoiced character since the energy in the upper frequency range increases for unvoiced utterances and vice versa. In principal all of the above presented energy based features are possible candidates for this feature. The choice of the highpass and lowpass filters is crucial for the reliability of the feature. Formally we can notate

$$r_{\mathrm{hlr}}(m) = \frac{W_{\mathrm{high}}(m)}{W_{\mathrm{low}}(m)}, \tag{3.117}$$

where $W_{\mathrm{high}}(m)$ and $W_{\mathrm{low}}(m)$ denote the energy of a highpass filtered and a lowpass filtered signal, respectively, with arbitrary cut-off frequencies. It is very important to retain the same type of energy feature used and the same highpass and lowpass filters during a training phase and the operation phase of the approaches presented in Chap. 5 since the features depend crucially on these parameters.

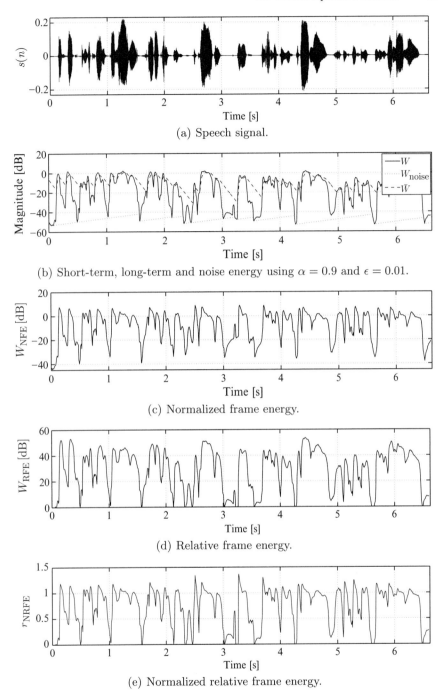

(a) Speech signal.

(b) Short-term, long-term and noise energy using $\alpha = 0.9$ and $\epsilon = 0.01$.

(c) Normalized frame energy.

(d) Relative frame energy.

(e) Normalized relative frame energy.

Fig. 3.15. Examples for several energy based features

In this section some energy based features have been presented. The only feature that has been used for classification tasks within this work is the high-pass energy to lowpass energy ratio. It has been used for a voiced/unvoiced classification of the current utterance (see Sect. 5.3.2). The relative frame energy or SNR has been employed within the control part of a system for bandwidth extension to switch off the extension if a certain threshold has been exceeded to avoid malclassifications for the spectral envelope estimation and thereby bothersome artifacts.

3.4 Distance Measures

Distance measures play a very important role in speech quality assessment, codebook training for speech coding or bandwidth extension and codebook search. Plenty of different distance measures for all kinds of applications exist. Since we are dealing with bandwidth extension and one major task is the extension of the spectral envelope, we will focus on distance measures that are appropriate to evaluate distances between spectral envelopes.

Mainly two different types of distance measures are common in digital signal processing. These are defined in the time domain and in the frequency domain. The most popular choice for both approaches is the use of the mean squared error. For bandwidth extension algorithms however either distance measures making use of the parametric representation of the spectral envelope directly are employed or measures making use of the spectrum that is derived from a parametric representation of the spectral envelope. Most of the spectral distance measures are L_p-norm based measures

$$d_p\left(S, \hat{S}\right) = \left[\frac{1}{2\pi} \int_{-\pi}^{\pi} \left|S(e^{j\Omega}) - \hat{S}(e^{j\Omega})\right|^p d\Omega\right]^{1/p}. \qquad (3.118)$$

The most common choices for p are $p = 1, 2, \infty$ resulting in the so-called *city block distance, Euclidean distance* and *Minkowski distance*. All L_p-norm based measures satisfy the constraints of a metric since they are symmetric and fulfill the triangular inequality.

3.4.1 Log Spectral Deviation

Since the representation of the magnitude spectrum in a logarithmic manner using the dB domain is very popular, another well known distance measure has emerged from this kind of representation. The *log spectral deviation* makes use of the city block distance

$$d_{\mathrm{LSD}}\left(S, \hat{S}\right) = \frac{1}{2\pi} \int_{-\pi}^{\pi} \left|20 \log_{10}\left|\frac{S(e^{j\Omega})}{\hat{S}(e^{j\Omega})}\right|\right| d\Omega \qquad (3.119)$$

$$= d_1\left(20 \log_{10}(S), 20 \log_{10}(\hat{S})\right).$$

As already mentioned above, the most interesting application in terms of bandwidth extension for distance measures is the comparison of spectral envelopes. Therefore $S(e^{j\Omega_k})$ can be replaced by the spectrum of the vocal tract filter $1/A(e^{j\Omega_k})$ which is easily derived from the predictor coefficients a_i by zero padding and a subsequent DFT/FFT.

3.4.2 RMS-Log Spectral Deviation

The root mean square (RMS) log spectral deviation is based on the Euclidean distance

$$d_{\text{RMS-LSD}}^2\left(S,\hat{S}\right) = \frac{1}{2\pi}\int\limits_{-\pi}^{\pi}\left|20\log_{10}\left|\frac{S(e^{j\Omega})}{\hat{S}(e^{j\Omega})}\right|\right|^2 d\Omega \qquad (3.120)$$

$$= d_2^2\left(20\log_{10}(S), 20\log_{10}(\hat{S})\right).$$

3.4.3 Cepstral Distance

A distance measure that is computed directly on the parametric representation of the spectral envelope using cepstral coefficients is the so-called *cepstral distance*. The cepstral distance is defined as

$$d_{\text{ceps}}\left(\mathbf{c},\hat{\mathbf{c}}\right) = \sum_{k=-\infty}^{\infty}\left(c_k - \hat{c}_k\right)^2, \qquad (3.121)$$

where the c_k denote the cepstral coefficients as shown in Sect. 3.2.3. An interesting property of this definition is its proportionality to the RMS-log spectral deviation from Sect. 3.4.2. This can be shown using Parseval's theorem [Rabiner 93]

$$d_{\text{ceps}}\left(\mathbf{c},\hat{\mathbf{c}}\right) = \sum_{k=-\infty}^{\infty}\left(c_k - \hat{c}_k\right)^2$$

$$= \frac{1}{2\pi}\int\limits_{-\pi}^{\pi}\left|\ln\left|\frac{S(e^{j\Omega})}{\hat{S}(e^{j\Omega})}\right|\right|^2 d\Omega$$

$$= \left(\frac{1}{20\log_{10}e}\right)d_{RMS-LSD}^2\left(S,\hat{S}\right)$$

$$= \left(\frac{1}{20\log_{10}e}\right)d_2^2\left(20\log_{10}(S), 20\log_{10}(\hat{S})\right). \qquad (3.122)$$

For a finite amount of cepstral coefficients the sum reduces to

$$d_{\text{ceps}}\left(\mathbf{c},\hat{\mathbf{c}}\right) = \sum_{k=1}^{N_{\text{ceps}}}\left(c_k - \hat{c}_k\right)^2. \qquad (3.123)$$

The sum in (3.123) starts at $k = 1$ since the coefficient c_0 only contains information on the power of the spectrum and for cepstral coefficients that are derived from predictor coefficients (see (3.84)) c_0 is equal to zero anyway.

3.4.4 Likelihood Ratio Distance

Another distance measure that is advantageous for the use with the parametric representation of the spectral envelope using LPC coefficients is the *likelihood ratio distance*. The likelihood ratio distance, replacing the speech spectrum $S\left(e^{j\Omega}\right)$ by the spectral envelope $1/A\left(e^{j\Omega}\right)$ in the following, is defined as

$$d_{\mathrm{LR}}\left(A, \hat{A}\right) = \frac{1}{2\pi} \int_{-\pi}^{\pi} \frac{\left|\hat{A}(e^{j\Omega})\right|^2}{\left|A(e^{j\Omega})\right|^2} \, d\Omega \; - 1. \tag{3.124}$$

Equation (3.124) can be computed very efficiently in discrete representation using only the vector $\mathbf{a}$ containing the predictor coefficients and the matrix $\mathbf{R}_{ss}$ with elements as described in Sect. 3.1.1.

$$d_{\mathrm{LR}}\left(\mathbf{a}, \hat{\mathbf{a}}\right) = \frac{\hat{\mathbf{a}}^{\mathrm{T}} \mathbf{R}_{ss} \, \hat{\mathbf{a}}}{\mathbf{a}^{\mathrm{T}} \mathbf{R}_{ss} \, \mathbf{a}} - 1. \tag{3.125}$$

Within this work this distance measure has been used for codebooks that have been trained on predictor coefficients directly as will be explained in Sect. 5.3.1. Since only the index i corresponding to the codebook entry producing the minimum of all distances (and not the minimum distance itself) is needed within a codebook search, it is sufficient to evaluate

$$\tilde{d}_{\mathrm{LR}}(n, i) = \hat{\mathbf{a}}_i^T \mathbf{R}_{ss}(n)\hat{\mathbf{a}}_i. \tag{3.126}$$

Note that (3.126) can be computed very efficiently since the autocorrelation matrix $\mathbf{R}_{ss}(n)$ has Toeplitz structure. However the likelihood ratio distance is not symmetric and therefore does not satisfy the constraints of a metric [Rabiner 93]. Beside the cost function according to (3.124), which weights the difference between the squared transfer function in a linear manner, a variety of distance measures that are similar to the likelihood ratio distance can be applied [Gray 80]. Most of them apply a spectral weighting function within the integration over the normalized frequency Ω. Furthermore, the difference between the spectral envelopes is often weighted in a logarithmic manner (instead of a linear or quadratic manner). The logarithmic approach takes the human loudness perception in a better way into account.

3.4.5 Itakura Distance

Another distance measure that is very popular has been defined by Itakura [Rabiner 93]

$$d_I\left(A, \hat{A}\right) = \ln \left[\frac{1}{2\pi} \int\limits_{-\pi}^{\pi} \frac{|\hat{A}(e^{j\Omega})|^2}{|A(e^{j\Omega})|^2} d\Omega \right] \tag{3.127}$$

$$= \ln \left[d_{LR}\left(A, \hat{A}\right) + 1 \right]. \tag{3.128}$$

3.4.6 Itakura–Saito Distance

The following distance measure was first published by Itakura and Saito and is one of the most popular distance measures in speech signal processing [Rabiner 93]

$$d_{IS}\left(A, \hat{A}\right) = \frac{1}{2\pi} \int\limits_{-\pi}^{\pi} \left[\frac{|\hat{A}(e^{j\Omega})|^2}{|A(e^{j\Omega})|^2} - \ln \frac{|\hat{A}(e^{j\Omega})|^2}{|A(e^{j\Omega})|^2} - 1 \right] d\Omega. \tag{3.129}$$

However it should be noted that, in parallel to the Itakura distance, the Itakura–Saito distance is not symmetric and therefore does not satisfy the constraints of a metric.

3.4.7 Other Spectral Distance Measures

The spectral distortion measures presented above do not take the human auditory system into account and do not penalize spectral overestimation higher than underestimation due to their symmetry [Nilsson 01]. In the following we will derive a own simple spectral distortion measure that takes the characteristic of the human auditory system into account. For deriving such a distance measure we first define the difference between the squared absolute values of the estimated and the original broadband spectra of the current frame as

$$\Delta\left(e^{j\Omega}\right) = 20 \log_{10} \left[\frac{|A\left(e^{j\Omega}\right)|}{|\hat{A}\left(e^{j\Omega}\right)|} \right]. \tag{3.130}$$

One basic characteristic of human perception is that with increasing frequency the resolution decreases. This fact is taken into account by adding an exponential decay of a weighting factor for increasing frequency. Another basic characteristic is, that if the magnitude of the estimated spectrum is above the magnitude of the original one, bothersome artifacts will occur. In the other case the estimated spectrum has less magnitude than the original one. This does not lead to artifacts that are as bothersome as the ones that occur when the magnitude of the estimated signal is above the original one. This characteristic implies the use of a non-symmetric distortion measure which we simply call *spectral distortion measure* (SDM), first published in [Iser 05b]

$$d_{SDM}\left(A, \hat{A}\right) = \frac{1}{2\pi} \int\limits_{-\pi}^{\pi} \xi\left(e^{j\Omega}\right) d\Omega, \tag{3.131}$$

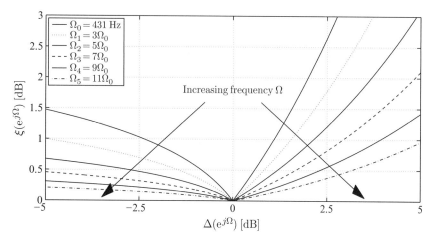

Fig. 3.16. Frequency dependent branches of d_{SDM} for $\alpha = 0.1$ and $\beta = 5$.

where $\xi(\mathrm{e}^{j\Omega})$ is implementing the perceptually motivated weighting of the logarithmic difference of the spectra and is therefore defined as

$$\xi\left(\mathrm{e}^{j\Omega}\right) = \begin{cases} \Delta\left(\mathrm{e}^{j\Omega}\right) \mathrm{e}^{\alpha\Delta\left(\mathrm{e}^{j\Omega}\right)-\beta\Omega}, & \text{for } \Delta\left(\mathrm{e}^{j\Omega}\right) \geq 0, \\ \ln\left[-\Delta\left(\mathrm{e}^{j\Omega}\right)+1\right]\mathrm{e}^{-\beta\Omega}, & \text{otherwise.} \end{cases} \tag{3.132}$$

Fig. 3.16 shows the characteristic of the above defined distortion measure for $\alpha = 0.1$ and $\beta = 5$. Note that this measure is unsymmetric and therefore does not satisfy the constraints of a metric.

As we will see in Sect. 6.1.3 this simple approach is able to outperform established distance measures when comparing the correlation of the results of an informal listening test to the results that different distance measures produce.

Concerning bandwidth extension the above presented distance measures are applied for evaluating the quality of different approaches as well as for classification tasks during the codebook search and have therefore a major impact on the quality of the extended speech signal. Distance measures that reflect the human perception better than the measures presented in this book involve higher computational complexity and have therefore not been subject of this study.

Within this work the cepstral distance measure has been employed for codebook training and search in the later application that are based on cepstral coefficients derived from predictor coefficients according to (3.84). The cepstral distance measure has also been used implicit within the learning algorithm for neural network training (see Sect. 5.4).

4

Excitation Signal Extension

This chapter deals with the estimation of the narrowband excitation signal, its extension and with finding an adequate gain factor for performing a power adjustment. Concerning bandwidth extension algorithms this represents one of two main parts as explained in Sect. 2.3. If we consider the source-filter model introduced in Sect. 2.2 the extension of the excitation signal corresponds to the process within the source path. To be more precise we are dealing with the three blocks *envelope extraction, excitation signal extraction* and *excitation signal extension* in Fig. 2.8. Figure 4.1 shows a more detailed view of these functional blocks. In the first processing stage the narrowband excitation signal $e_{(\text{nb})}(n)$ is extracted by the application of a whitening filter (predictor-error filter). This whitening filter depends on the LPC analysis of the narrowband input signal. The next functional block represents the main task within the extension of the excitation signal. Therefore we will present three major groups of methods for extending the excitation signal. They can further be distinguished in time-domain and frequency-domain approaches. The first group will be *non-linear characteristics* [Carl 94a, Carl 94b, Kornagel 03] which are applied in the time domain. The second one comprises so-called *spectral shifting approaches* (see [Kornagel 01], [Fuemmeler 01]), which can be applied in the time domain as well as in the frequency domain, followed by the third group, the so-called *function generators* that are most often applied in the time domain. For the latter we will only present the approach where the excitation signal is modeled by white noise or sine generators [Miet 00]. Another method which is not presented in this book due to its limited results concerning speech quality is the so called *fricative spreading* where the upper range of the spectrum is spread towards the extension region if a fricative has been detected [Heide 98],[Mason 00]. The next functional block in Fig. 4.1 performs an optional spectral whitening. This is due to the fact that some of the algorithms for extending the excitation signal presented in this book perform a spectral coloration which has to be reversed. In this chapter we also describe how to adjust the power of the estimated broadband excitation signal to the

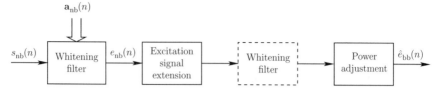

Fig. 4.1. Overall system for excitation signal extension

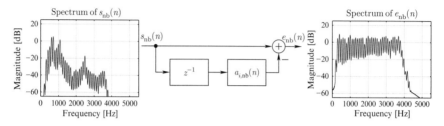

Fig. 4.2. Excitation signal extraction. On the left side of the predictor-error filter a short-term spectrum of the narrowband input speech signal is depicted and on the right side the respective short-term spectrum of the whitened signal

narrowband input which is represented by the last functional block in Fig. 4.1. This is a very important part since a wrong amplification or a temporal jitter in the amplification produces strong artifacts.

A detailed evaluation, including subjective as well as several objective criteria, of the different algorithms that are presented in this chapter can be found in Sect. 6.1.

4.1 Estimation of the Narrowband Excitation Signal

Before we are able to extend the narrowband excitation signal we first have to extract or estimate it from the narrowband speech signal. To achieve this we need to remove the spectral envelope. As shown in Sect. 3.1 this may be done by applying a low order predictor-error filter (see Fig. 4.2). Figure 4.2 realizes the two blocks *envelope extraction* and *excitation signal extraction* in Fig. 2.8. In Fig. 4.1 the filter coefficients are set externally and the predictor-error filter is described as a whitening filter. By removing the envelope using a predictor-error filter we can accomplish the task of estimating the narrowband spectral envelope as well. This estimate is needed in the filter path of Fig. 2.8. The estimated narrowband excitation signal $e_{nb}(n)$ is available in the time domain. Some of the algorithms presented in the following are frequency domain algorithms. Therefore, a DFT/FFT is applied for these algorithms before further processing.

However, the method that has been implemented in this work is slightly different from the above description. For this work the spectral envelope of the narrowband input signal has been determined but the estimate for the narrowband excitation signal has been derived applying a predictor-error filter driven by the coefficients of the respective broadband codebook entry (if available). This is the reason why the coefficients of the whitening filter in Fig. 4.1 are set externally. This has the advantage that the noise within the extension regions is not amplified as can be seen in Fig. 4.2 on the right hand side. Furthermore the needed high amplification for pushing the noise within the extension regions causes the filter to reproduce the spectral envelope in the passband in an insufficient manner.

4.2 Extension Using Non-Linear Characteristics

In this section we will present several non-linear characteristics which are appropriate for extending the narrowband excitation signal (see Fig. 4.3). These are applied in the time domain in this book. It is well known that applying a non-linear characteristic to a harmonic signal produces sub- and super-harmonics. The principle of this property can be shown using a quadratic characteristic as an example. The application of a quadratic characteristic in the time domain corresponds to a convolution in the frequency domain

$$\hat{e}_{\mathrm{bb}}(n) = e_{\mathrm{nb}}^2(n) \;\circ\!\!-\!\!\bullet\; E_{\mathrm{nb}}(e^{j\Omega_k}) * E_{\mathrm{nb}}(e^{j\Omega_k}) \tag{4.1}$$

$$= \sum_{i=-\infty}^{\infty} E_{\mathrm{nb}}(e^{j\Omega_i}) E_{\mathrm{nb}}(e^{j\Omega_{k-i}}) \tag{4.2}$$

$$= \hat{E}_{\mathrm{bb}}(e^{j\Omega_k}). \tag{4.3}$$

If we now consider a harmonic signal, as it would be the case with voiced speech segments, this convolution in the frequency domain leads to two line spectra which are shifted by Ω_k (see Fig. 4.4a). Each time the maxima of the line spectra coincide again (Ω_k is equal to an integer multiple of the pitch frequency Ω_0) a maximum within the resulting signal $\hat{E}_{\mathrm{bb}}(e^{j\Omega_k})$ is formed (see Fig. 4.4b). The different non-linear characteristics that are investigated in this book yield different properties concerning the production of even and/or odd harmonics (see Fig. 4.5). For example, a quadratic characteristic produces only even harmonics whereas a cubic one produces only odd harmonics. By applying the non-linear characteristic to a harmonic signal consisting of several even and odd harmonics this property does not affect the result seriously. However, there are few effects that have to be mentioned. First, a potentially resulting component at $\Omega_k = 0$ (see Figs. 4.6 and 4.5) has to be removed after the application of the non-linear characteristic. Another effect that might occur is aliasing (see Fig. 4.6). This can be circumvented by oversampling by an appropriate factor, then applying the non-linear characteristic, appropriate lowpass filtering, and finally downsampling again. Most non-linear characteristics perform a coloration of the originally white excitation signal. If this is

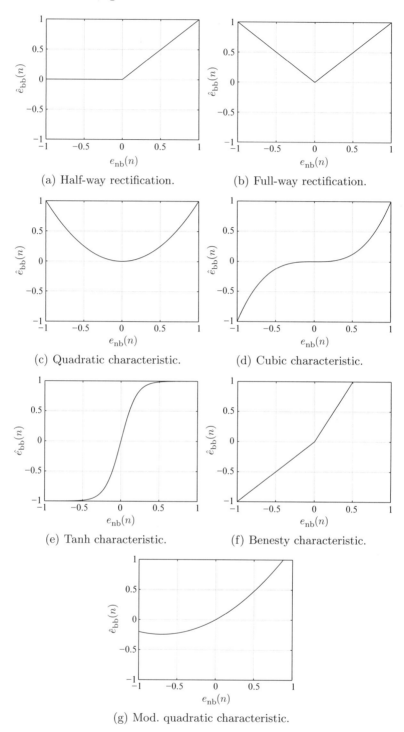

(a) Half-way rectification.

(b) Full-way rectification.

(c) Quadratic characteristic.

(d) Cubic characteristic.

(e) Tanh characteristic.

(f) Benesty characteristic.

(g) Mod. quadratic characteristic.

Fig. 4.3. Characteristic of different non-linear functions

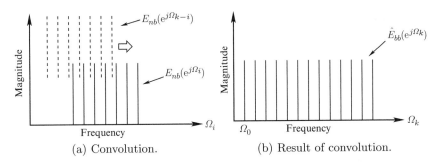

(a) Convolution.   (b) Result of convolution.

Fig. 4.4. (**a**) and (**b**) illustrate the effect of the convolution in the frequency domain. Note that this is only a schematic illustration. The resulting *line spectrum* typically exhibits a coloration due to the respective non-linear characteristic that has been applied and due to the aliasing that occurs depending on the sampling rate and the effective bandwidth

unintentional we can simply once again apply a predictor-error filter as described in Sect. 4.1. As we will see later on, when discussing other methods for the extension of the excitation signal, the big advantage of applying non-linear characteristics is the production of well placed harmonics. All other pitch adaptive extension algorithms depend crucially on the accuracy and reliability of the applied pitch detection algorithm (see Sect. 3.3.3).

4.2.1 Half-Way Rectification

A very well known non-linear characteristic originates from the field of classical electrical engineering. As indicated by the name this characteristic has been used to rectify alternating current by blocking the negative half wave and letting pass the positive one. Figure 4.3a illustrates this behavior. Formally we can describe this phenomenon as

$$\hat{e}_{\mathrm{bb}}(n) = \begin{cases} 0, & \text{for } e_{\mathrm{nb}}(n) \leq 0, \\ e_{\mathrm{nb}}(n), & \text{otherwise.} \end{cases} \tag{4.4}$$

One property worth mentioning is that after processing with a half-way rectifier the signal is no longer zero mean. Another property is that the half way rectifier is not power conserving. The half way rectifier produces even harmonics including the fundamental frequency (see Fig. 4.5a).

4.2.2 Full-Way Rectification

Like the half-way rectifier the full-way rectifier springs from the analog world. The difference in the functionality lies in the reversion of the negative half wave

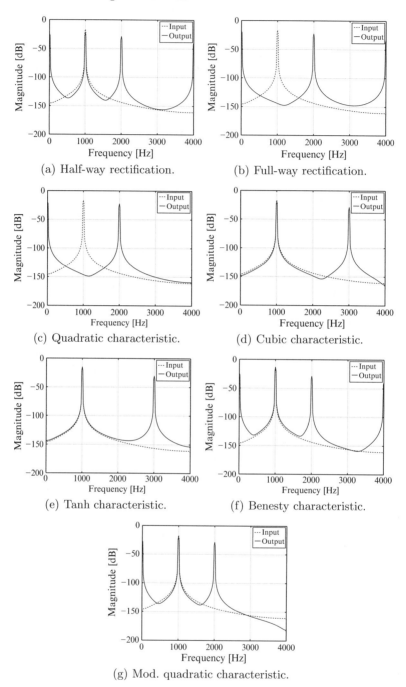

(a) Half-way rectification.

(b) Full-way rectification.

(c) Quadratic characteristic.

(d) Cubic characteristic.

(e) Tanh characteristic.

(f) Benesty characteristic.

(g) Mod. quadratic characteristic.

Fig. 4.5. Effect of applying different non-linear characteristics to a 1 kHz sine tone. Although the input signal comprises only one discrete frequency, components in between are visible due to windowing and numerical effects

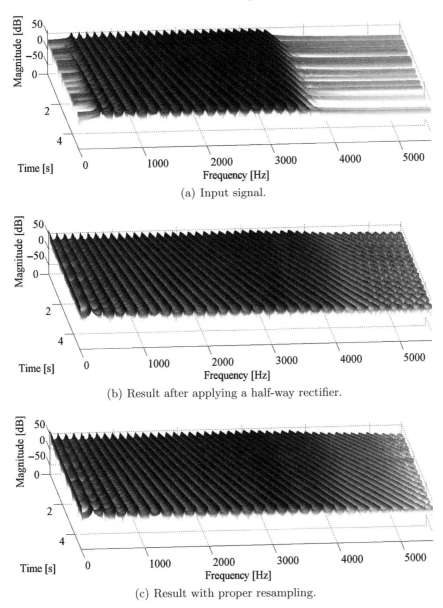

(a) Input signal.

(b) Result after applying a half-way rectifier.

(c) Result with proper resampling.

Fig. 4.6. Effect of the application of a half-way rectifier to a harmonic signal with increasing fundamental frequency which is bandlimited. Note the resulting constant component in (**a**) and (**b**) and the aliasing in (**b**) that occurs without proper resampling and filtering. Although the input signal comprises discrete frequencies, components in between are visible due to windowing and numerical effects

considering the rectification of alternating current. In Fig. 4.3b the behavior of the full way rectifier is depicted. The function can be written as

$$\hat{e}_{bb}(n) = \left| e_{nb}(n) \right|. \tag{4.5}$$

Just as the half-way rectifier the full-way rectifier produces an output signal which is no longer zero mean. But in contrast to the half way rectifier the full way rectifier is power conserving. The full way rectifier produces even harmonics without the fundamental frequency (see Fig. 4.5b).

4.2.3 Quadratic Characteristic

Another well known non-linear function is the quadratic function. Figure 4.3c shows this function. The function is characterized by

$$\hat{e}_{bb}(n) = e_{nb}^2(n). \tag{4.6}$$

As well as the already presented characteristics, the quadratic function produces an output signal that is not zero mean. Additionally the power of the signal is changed. The solely production of even harmonics produces a very pleasant sounding signal. The quadratic characteristic produces the second harmonic without the fundamental frequency (see Fig. 4.5c).

4.2.4 Cubic Characteristic

In contrast to the quadratic characteristic, the cubic characteristic produces odd harmonics. The function is defined by

$$\hat{e}_{bb}(n) = e_{nb}^3(n). \tag{4.7}$$

In Fig. 4.3d a plot of the function is depicted. The output of the function is zero mean presuming that the input sequence is zero mean and possesses a symmetric density distribution. The function is not power conserving. Even more than by using the quadratic characteristic, the dynamics of the output signal have to be observed and limited if necessary. The cubic characteristic produces the third harmonic including the fundamental frequency (see Fig. 4.5d).

4.2.5 Tanh Characteristic

A function that is used because of its similarity to saturation characteristics that occur for example by operating a vacuum tube near the distortion limit is the hyperbolic tangent

$$\hat{e}_{bb}(n) = \tanh \left(\mu e_{nb}(n) \right). \tag{4.8}$$

This function is depicted in Fig. 4.3e for $\mu = 3$. The output is zero mean presuming that the input is zero mean and has a symmetric density distribution but the function is not power conserving. The tanh characteristic produces odd harmonics including the fundamental frequency (see Fig. 4.5e).

4.2.6 Benesty Characteristic

Another function that has been investigated in this book originates from the field of acoustic echo cancellation. The characteristic is used to decorrelate stereo signals [Benesty 01]. The characteristic is defined by

$$\hat{e}_{bb}(n) = e_{nb}(n) + \alpha \frac{e_{nb}(n) + |e_{nb}(n)|}{2}. \tag{4.9}$$

In Fig. 4.3f the characteristic is shown for $\alpha = 1$. In this configuration the characteristic is not power conserving and its output is no longer zero mean assuming a zero mean input signal with symmetric density distribution. The Benesty characteristic produces even harmonics including the fundamental frequency (see Fig. 4.5f).

4.2.7 Adaptive Quadratic Characteristic

To circumvent the problem of distinct dynamics using a quadratic characteristic but maintaining the pleasant sound of the quadratic characteristic, an adaptive limiting mechanism has been introduced. Additionally a mixture of a quadratic characteristic and a half-way rectification has been attempted to be implemented since the half way rectification shows convincing results as well (see Sect. 6.1). This characteristic has been derived during this work and has first been published in [Iser 06].

The adaptive quadratic characteristic uses a linear as well as a quadratic term and is defined as

$$\hat{e}_{bb}(n) = c_2(n)e_{nb}^2(n) + c_1(n)e_{nb}(n) + c_0(n). \tag{4.10}$$

Aim of the characteristic is to limit the maximum and minimum output $(\hat{e}_{bb,max}(n), \hat{e}_{bb,max}(n))$ by a factor

$$\hat{e}_{bb,max}(n) = k_1 e_{nb,max}(n), \tag{4.11}$$

$$\hat{e}_{bb,min}(n) = -k_2 e_{nb,max}(n), \tag{4.12}$$

where $e_{nb,\,max}(n)$ describes the smoothed short-term magnitude maximum of $e_{nb}(n)$. If we use these constraints in (4.10) and set $c_0 = 0$, forcing the characteristic to go through the origin, we can solve for the coefficients c_2 and c_1

$$c_2(n) = \frac{k_1 - k_2}{2\,e_{nb,max}(n) + \epsilon}, \tag{4.13}$$

$$c_1(n) = \frac{k_1 + k_2}{2}. \tag{4.14}$$

The coefficient $c_0(n)$ can be used for compensating the resulting constant component. For avoiding divisions by zero a small number $\epsilon > 0$ is added in the denominator of (4.13). The characteristic of this function is depicted in Fig. 4.3g for $k_1 = 1.2$, $k_2 = 0.2$ and $e_{nb, \, max}(n) = 1$. This proved to be a good choice for k_1 and k_2 since we do want to preserve the power of the input signal and still cause non-linear behavior. The adaptive quadratic characteristic produces the second harmonic including the fundamental frequency (see Fig. 4.5g). The major advantages of this adaptive approach is the fact that clipping can be avoided since the amplitude of the input signal is tracked. Another aspect of the adaptive implementation is the fact that low energy as well as high energy input signals are both processed in the same non-linear manner. For fixed non-linear functions as for example the tanh characteristic low energy input signals are handled in a linear manner whereas high energy input signals are handled in a highly non-linear manner.

As we will see later on, the adaptive quadratic characteristic outperforms the other approaches particulate significantly.

4.3 Extension Using Spectral Shifting/Modulation Techniques

The basic idea behind these approaches is to exploit the presence of an existing harmonic structure in the telephone band assuming voiced utterances and shifting a copy in different manners into the excitation regions.

Modulation techniques is a term that implies the processing of the excitation signal in the time domain by performing a multiplication

$$\hat{e}_{bb}(n) = e_{nb}(n)e^{jn\Omega_0} \circ\!\!-\!\!\bullet E_{nb}\left(e^{j\Omega}\right) * \delta(\Omega - \Omega_0) = \hat{E}_{bb}\left(e^{j\Omega}\right). \qquad (4.15)$$

This multiplication in the time domain corresponds to the convolution with a dirac function in the frequency domain resulting in a shift. This shift can be performed equivalently in the frequency domain

$$\hat{E}_{bb}\left(e^{j\Omega}\right) = E_{nb}\left(e^{j(\Omega-\Omega_0)}\right) \qquad (4.16)$$

hence the corresponding frequency domain operations are known under the term spectral shifting.

4.3.1 Fixed Spectral Shifting

Fixed spectral shifting indicates a shift with a fixed frequency offset. In Fig. 4.7 the frequency domain approach is depicted schematically. The starting index of the copy for the upper and lower extension region can be chosen independently. However, when choosing the starting index for the copy of the upper extension region one has to keep in mind not to exceed the upper band limit given by the telephone bandpass, nor the Nyquist limit. The so derived signal $\hat{e}_{bb}(n)$ or $\hat{E}_{bb}\left(e^{j\Omega}\right)$, respectively is then filtered with the complementary

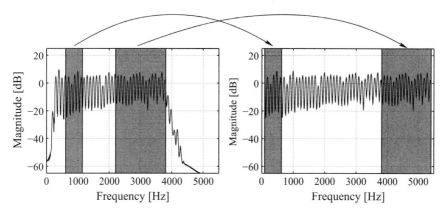

Fig. 4.7. Spectral shifting

bandstop corresponding to the telephone bandpass and added to the original narrowband excitation signal. The reason for performing such a shift can easily be understood when we consider the processing of voiced utterances. The periodic signal that is typical for voiced utterances can be extended by these modulation techniques. As seen in Sect. 2.1 the excitation signal of unvoiced utterances is noise-like and therefore no need to take care of a specific structure exists. This means that we can also apply this excitation extension method for unvoiced utterances but without having to worry about discontinuing a structure. The possibility of continuing a structure in an erroneous manner using a fixed shifting scheme in the case of voiced utterances and the small noticeable artifacts affiliated, lead to a pitch-adaptive implementation presented in the following. A disadvantage that has to be mentioned is the fact that the phase information that is also copied into the extension regions might not be correct and could therefore lead to audible artifacts.

4.3.2 Adaptive Spectral Shifting

Incorporating a pitch detection and estimation like the algorithm presented in Sect. 3.3.3 one could even perform an adaptive shift and thereby keep the right pitch structure even at the transition regions from the telephone bandpass to the extension region [Fuemmeler 01, Gustafsson 01]. This can be done choosing the starting index for the copy such that the gap between the last harmonic component in the telephone band and the first one in the copy just equals the fundamental frequency that is known through the pitch determination algorithm.

Experiments using a pitch adaptive spectral shifting showed that the speech quality critically depends on the quality of the pitch determination algorithm. Jitter in the pitch detection results in audible artifacts. However the improvement that can be achieved by an adaptive shift concerning the upper extension region is only marginal.

4.4 Extension Using Function Generators

Function generators in principle work in the time domain. In the following we will present two types of function generators:

- The first one, the so-called *sine generators*, are intended to extend the excitation signal of voiced utterances
- whereas the second one, the so-called *white noise generators*, are intended to extend the excitation signal of unvoiced utterances

Since the human auditory system is not very sensitive to pitch misplacement or missing pitch in the upper extension region it is advantageous to combine these two methods by using the sine generators for the lower extension region and the white noise generator for the upper one. The benefit of using the sample based time domain function generators is the absence of problems with phase discontinuity as arising by using a block based frequency domain algorithm.

4.4.1 Sine Generators

The first class of algorithms for excitation signal extension using function generators are sine generators. The simplest form of a function generator is a sine generator. Similar to the use of adaptive modulation techniques this method needs a pitch estimation. The sine generators are working in the time domain. The parameters (amplitude, frequency) of the sine generators are obtained by employing the estimated broadband spectral envelope for determining the appropriate amplitude if not a white signal is needed and the estimate for the fundamental frequency and its harmonics, respectively. For the task of excitation signal extension several sine generators may be necessary in order to produce a harmonic spectrum. The advantage of the use of sine generators lies in the differentiation between actual values for amplitude and frequency and the desired values for amplitude and frequency. The sine generators are designed to change their current parameters such as amplitude and frequency smoothly within a maximum allowed change of the parameters towards the desired values. This prevents artifacts due to a step of amplitude or frequency from one frame to another. Another advantage of these sine generators is that artifacts due to a step in phase of the low frequency components do not appear due to the time domain processing. Furthermore, the sine generators do not need an estimated value for the fundamental frequency with every sample or frame but for example whenever such an estimate is very reliable.

4.4.2 White Noise Generators

A second class of function generators comprises so-called *white noise generators* [Qian 03], [Qian 04a], [Taori 00]. This kind of function generators are preferably used to extend the excitation signal of unvoiced utterances. White noise generators yield the advantage that no pitch information is needed. But for the lower extension region they are inappropriate since unvoiced speech segments have low power in the lower extension region and if applied to voiced speech segments in the lower extension region bothersome artifacts are produced. The white noise generators in this book have been implemented using pseudo-noise sequences (PN sequences) of appropriate length produced by a back coupled shifting register [Hänsler 04]. This approach produces values $\in \{0,1\}$. These values are used as bit patterns to produce a random 16 bit integer number. Approaches testing for example other amplitude distributions that are more similar to the amplitude distribution of speech than uniform or Gaussian distributions may increase the quality of this group of extension methods.

4.5 Power Adjustment

For algorithms using one of the above presented non-linear characteristics that are not power conserving, it is necessary to adjust the power of the extended excitation signal. Especially the output of a quadratic or cubic characteristic which is highly dynamic needs to be adjusted in power. By whitening the input signal using a predictor-error filter and thereby extracting the narrowband excitation signal as well as the spectral envelope of the input signal, we also derived the power of the residual error signal implicitly by applying the Levinson–Durbin recursion (see Appendix A). Now we have to determine the power of the extended excitation signal within the telephone band. This can be realized in the frequency domain by limiting the integral to the frequencies corresponding to the telephone band

$$W_{\text{ext,tb}} = \int_{\Omega_{\text{low}}}^{\Omega_{\text{high}}} \hat{E}_{\text{bb}}^2 \left(e^{j\Omega} \right) \mathrm{d}\Omega. \tag{4.17}$$

The final gain factor for adjusting the power of the extended excitation signal can then be computed as the ratio of these two powers

$$g_{\text{ext}} = \sqrt{\frac{W_{\text{nb}}}{W_{\text{ext,tb}}}}, \tag{4.18}$$

where W_{nb} denotes the power of the narrowband excitation signal which is provided by the Levinson–Durbin recursion. Important for the validity of these operations is the assumption that the extended excitation signal is a white signal, which has to be guaranteed by applying an additional predictor-error filter, when needed.

4.6 Discussion

The most crucial issue in extending the excitation signal is the power adjustment. Small deviations in the power of the estimated broadband excitation signal from the real one are reflected in audible and bothersome artifacts. Mismatch in pitch for the upper extension region however does not effect the quality vitally. In contrast to this a pitch mismatch of the estimated broadband excitation signal in the lower extension region does interfere the speech quality drastically. Another issue for the lower extension region is the phase of the synthesized signal. Since the algorithm presented in this book is based on block processing it might happen that discontinuities in phase occur passing from one frame to another. This results in strong audible artifacts. The method presented in this chapter using sine generators is able to retain the phase from one frame to another but depends strongly in terms of quality on the robustness of the pitch detection.

Experiments using the original broadband spectral envelope and the different extension algorithms presented above have shown that the extension of the excitation signal is noncritical compared to the challenging task of extending the spectral envelope. However imprecisions sum up and combining two inaccurate parts makes the result even worse. Detailed analysis and the result of objective and subjective evaluation of the different algorithms are presented in Sect. 6.1.

In Sect. 6.1 not all methods presented above are evaluated. For the different parts (extending the upper or lower frequencies, respectively) a selection of algorithms has been examined. The methods rejected for further analysis have either been sorted out by performing a first informal subjective listening test and/or the immense additional effort needed without appreciable gain as it is the case for the pitch adaptive spectral shifting approach and the sine generators for example.

5

Broadband Spectral Envelope Estimation

One of the most demanding challenges in bandwidth extension algorithms based on the source-filter model, introduced in Sect. 2.2, is the estimation of the broadband spectral envelope as already mentioned in Sect. 2.3. All methods presented in this book for the estimation of the broadband spectral envelope are based on the narrowband spectral envelope as an input. Experiments have been conducted using additional scalar features and are described in the respective section [Jax 04b]. The extraction of the narrowband spectral envelope as well as the different methods for estimating the broadband spectral envelope will be matter of this chapter. For the extraction of the narrowband spectral envelope we will introduce a method that has been developed within this work for increased robustness against deviations in the telephone bandpass during operation leading to an improved classification. The methods for estimating the broadband spectral envelope out of the narrowband one incorporate codebooks, neural networks and some linear mapping approaches. Statistically motivated approaches using HMMs and GMMs can be found in [Jax 00, Jax 03b, Jax 03a, Park 00, Qian 04b, Yao 05]. Figure 5.1 shows the filter part of the general block diagram for BWE algorithms as introduced within Fig. 2.8 consisting of the extraction of the narrowband spectral envelope $\mathbf{a}_{\mathrm{nb}}(n)$ and the estimation of the broadband spectral envelope $\hat{\mathbf{a}}_{\mathrm{bb}}(n)$. Note that the extraction of scalar features is missing for simplicity. The chapter will close with a short discussion.

5.1 Generation and Preparation of a Training Data Set

Before being able to train a codebook, neural network, or linear mapping matrices, one important precondition is the availability of a sufficient amount of training data that has to be processed carefully to extract the needed features (see Sects. 3.2 and 3.3). Starting point is a broadband speech corpus with at least the desired target sampling rate and bandwidth, respectively. Such a

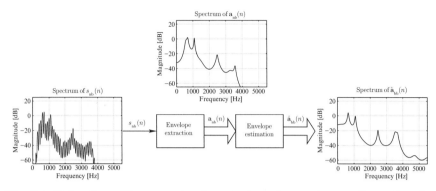

Fig. 5.1. Extraction of the narrowband spectral envelope and succeeding estimation of the broadband spectral envelope

speech corpus should meet the requirements of the later application as closely as possible concerning environmental noise level, the kind of utterances (read text or natural speech) and so on. Hence for the training of the approaches presented in this book an additional speech corpus has been recorded. This speech corpus consisted of recordings of 31 male and 16 female speakers including spontaneous speech as well as read texts. Another requirement from the car environment was the recording of *Lombard speech*. Lombard speech determines the effect that humans try to speak louder when the surrounding noise level is higher. Simultaneously the formants are shifted towards higher frequencies by speaking at a higher volume. The car noise has been simulated by using headphones playing a binaural recording of a car interior noise environment when driving with a speed of $100 \, \mathrm{km \, h^{-1}}$ at the appropriate sound pressure level. The recordings have been equalized for the microphone frequency response of the respective headset. The method used for doing so is presented in more detail later on.

This speech data can then be downsampled if needed. After removing the speech pauses the speech corpus then has to be processed in an appropriate manner to simulate a realistic band limiting transmission line. Another possibility is not to simulate such a behavior but to record the real transmitted data. The speech corpus in this book has been played back using a head and torso simulator (HATS) within an anechoic chamber (see Fig. 5.2a). The corpus has been equalized concerning the frequency response as well as the sound pressure level so that the signal at the mouth reference point (MRP) of the HATS equals the signal as it would have been recorded at the MRP of each person (see Fig. 5.3). For this task a test sequence $x(n)$ of white noise of sufficient length N has been generated and played back over the HATS. This signal has been recorded by a calibrated measurement microphone in the MRP (see Fig. 5.3a). If we denote the recorded signal with $y(n)$ we can estimate the frequency response of the HATS as

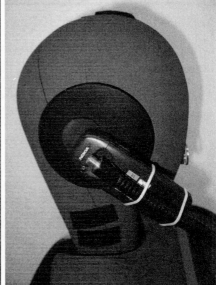

(a) HATS in anechoic chamber.   (b) Mobile phone attached to HATS in side view.

(c) Mobile phone attached to HATS in front view.

Fig. 5.2. Setup using a HATS and an anechoic chamber for playing back the speech corpus and succeeding transmission via GSM

$$H_{\text{HATS}}(e^{j\Omega}) \approx \frac{r_{yy}(0) + \sum\limits_{k=1}^{N} \left[r_{yy}(k)\, e^{-j\Omega k} + r_{yy}^{*}(k)\, e^{j\Omega k} \right]}{r_{xx}(0) + \sum\limits_{k=1}^{N} \left[r_{xx}(k)\, e^{-j\Omega k} + r_{xx}^{*}(k)\, e^{j\Omega k} \right]}. \tag{5.1}$$

This frequency response has been smoothed by applying an IIR-filter in frequency direction in forward as well as in backward operation. After inverting

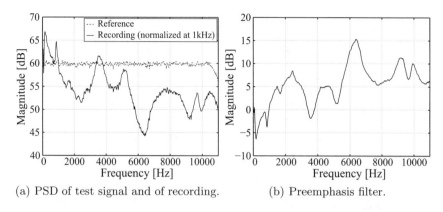

(a) PSD of test signal and of recording.   (b) Preemphasis filter.

Fig. 5.3. (a) PSD of the test signal used and the PSD of the test signal played over the HATS and recorded by a calibrated measurement microphone. (b) The frequency response of a preemphasis filter for equalizing the effect of the HATS

the frequency response we obtain the frequency response of an appropriate preemphasis filter for equalizing the frequency response of the HATS.

A Nokia 6310i mobile phone has been attached at the HATS (see Figs. 5.2b and 5.2c) and the far-end signal has been recorded after transmission over the german *D1-network* (GSM 900 MHz). As device on the far-end side a Siemens MC35i GSM modem has been used which was accessible through a serial port and an appropriate user interface.

After the band limitation the most crucial task concerning later quality is to compensate for the delay introduced by the band limiting components. The two data sets (narrowband and broadband) need to be synchronous. This synchronization has been performed in this book by filtering both, the broadband signal as well as the narrowband signal, by an additional bandpass with a passband that has to have the cut off frequencies within the passband of the bandpass, the narrowband signal has been processed with. The filtering has been performed in a forward and backward manner so that no additional delay has been introduced. After the filtering, a blockwise crosscorrelation has been computed between the narrowband and the former broadband signal. The offset of the position of the maximum of the crosscorrelation to the index 0 then indicates the delay which has been introduced into the narrowband signal by the processing done within the transmission. Additional variable delay introduced by so-called clock drift between the sound card used for playback and the one used for recording occurred but was negligible and has therefore not been compensated.

After ensuring that the signals are synchronous and they correlate sufficiently the features can be extracted. This has been done in a blockwise manner in this book. These features can comprise several parametric representations of the spectral envelope or any other interesting features (see

Sects. 3.2 and 3.3). If the transmission was not simulated but real, the correlation between the narrowband and the broadband features may get destroyed by drop outs or comfort noise injection during the transmission. Therefore signal blocks that did not correlate sufficiently concerning a quadratic distance measure comparing the spectral envelope representation within the passband, have been deleted. The remaining speech corpus consisted of $N_{\text{feat}} = 47{,}94{,}592$ features which equals an overall net length of the speech corpus of 116 min (as this does not look like a sufficient amount of speech data we should keep in mind, that all speech pauses have been removed).

The whole process of the needed training data preparation is depicted in Fig. 5.4.

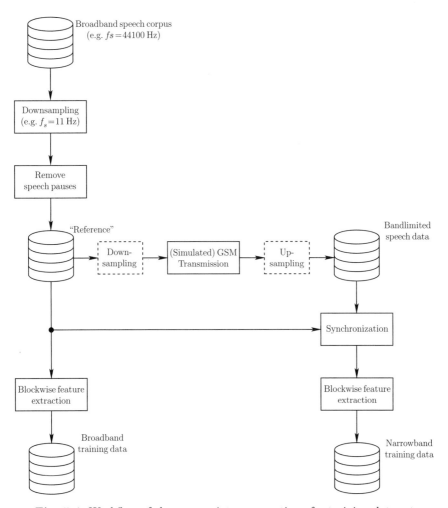

Fig. 5.4. Workflow of the appropriate preparation of a training data set

5.2 Estimation of the Narrowband Spectral Envelope

The estimation of the narrowband spectral envelope has been discussed briefly in Sect. 4.1. The extraction of the narrowband excitation signal can in general be accomplished in one step together with the estimation of the narrowband spectral envelope by simply applying a predictor-error filter as depicted in Fig. 2.8. The coefficients $a_{i,nb}(n)$ of the error filter depicted in Fig. 4.2 then represent the inverse of the narrowband spectral envelope. However, the coefficients obtained applying this method depend strongly on the particular implementation of the bandpass that has been used for transmission and on the quality of the sample rate converter used to convert the incoming 8 kHz signal to 11,025 Hz. The telephone bandpass may differ from the one used during the training phase. This can lead to malclassifications since the focus of the error criterion during the distance evaluation is also on searching the envelope that fits the slope at the cut off frequencies best and has the same stop band attenuation. But this is not of interest. Focus of the error criterion should be the shape of the envelope within the passband. Therefore the autocorrelation coefficients used within the computation of the predictor coefficients have been modified. Since all most often used types of coefficients can be derived using autocorrelation or linear predictive coefficients, the use of autocorrelation coefficients does not represent any limitation. The modification consists of the addition of *noise* that is spectrally shaped according to Fig. 5.5b which causes the signal outside the passband to be bound to that noise level. This is applied during the training phase as well as during the operation of the algorithm to the autocorrelation coefficients. It guarantees

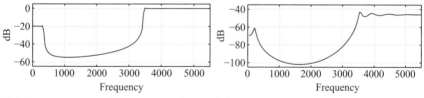

(a) Design using an higher order filter.   (b) Resulting spectrum of autocorrelation coefficients.

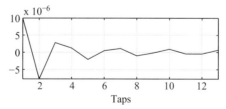

(c) Autocorrelation coefficients.

Fig. 5.5. Design of the additive perturbance for the autocorrelation coefficients

the envelopes to be almost identical outside the passband. Figure 5.5a shows the design of a *bandstop* using an FIR filter of order N_{FIR}

$$\mathbf{b} = [b_0, b_1, \ldots, b_{N_{\text{FIR}}-1}]^{\text{T}}. \tag{5.2}$$

The magnitude spectrum of the frequency response

$$B\left(e^{j\Omega_k}\right) = \mathcal{F}\left\{\mathbf{b}\right\}, \tag{5.3}$$

of this filter has been squared and transformed back into the time domain by an IDFT of order N_{DFT}

$$\mathbf{r} = \mathcal{F}^{-1}\left\{\left|B\left(e^{j\Omega_k}\right)\right|^2\right\}. \tag{5.4}$$

By doing so we obtain the autocorrelation coefficients

$$\mathbf{r} = [r_0, r_1, \ldots, r_{N_{\text{DFT}}-1}]^{\text{T}}, \tag{5.5}$$

of a signal with the according spectral shape. By truncating the autocorrelation sequence according to the order of the linear prediction and by introducing an additional attenuation

$$\mathbf{r}_{\text{mod}} = \mu \frac{\mathbf{W}_{\text{cut}}\mathbf{r}}{r_0}, \tag{5.6}$$

we obtain a signal with a spectral shape according to Fig. 5.5b with the respective autocorrelation coefficients depicted in Fig. 5.5c

$$\mathbf{r}_{\text{mod}} = \left[r_{\text{mod},0},\ r_{\text{mod},1}, \ldots, r_{\text{mod},N_{\text{lpc}}-1}\right]^{\text{T}}. \tag{5.7}$$

Where the $N_{\text{lpc}} \times N_{\text{DFT}}$ matrix $\mathbf{W}_{\text{cut}}$ has the following form:

$$\mathbf{W}_{\text{cut}} = \begin{bmatrix} w_{1,1} & 0 & \cdots & 0 & 0 \cdots 0 \\ 0 & w_{2,2} & \cdots & 0 & 0 \cdots 0 \\ \vdots & \vdots & \ddots & \vdots & \vdots \ \vdots \\ 0 & 0 & \cdots & w_{N_{\text{lpc}},N_{\text{lpc}}} & 0 \cdots 0 \end{bmatrix}. \tag{5.8}$$

Where the $w_{i,j}$ are defined as

$$w_{i,j} = 1, \text{ for } i = j \left\{j \in \mathbb{Z} \,\middle|\, j = [1, \ldots, N_{\text{lpc}}]\right\}. \tag{5.9}$$

The coefficient μ scales the amount of *noise* that is added as we will see later on. Here $\mu = 0.001$ has been used (differs from Fig. 5.5, where μ was set to $\mu = 0.00001$). For the order of the DFT and the linear predictive analysis we can formulate the following constraint:

$$N_{\text{DFT}} \geq N_{\text{lpc}}. \tag{5.10}$$

The so derived autocorrelation coefficients are now added to the normalized autocorrelation coefficients of the narrowband input signal

$$\mathbf{r}_{\mathrm{lpc,mod}}(n) = \frac{\mathbf{r}_{\mathrm{lpc}}(n) + \mathbf{r}_{\mathrm{mod}}}{1 + r_{\mathrm{mod},0}}. \tag{5.11}$$

Where the autocorrelation coefficients of the narrowband input signal

$$\tilde{\mathbf{r}}_{\mathrm{lpc}}(n) = [\tilde{r}_{\mathrm{lpc},0}(n),\ \tilde{r}_{\mathrm{lpc},1}(n),\ \ldots,\ \tilde{r}_{\mathrm{lpc},N_{\mathrm{lpc}}-1}(n)]^{\mathrm{T}}, \tag{5.12}$$

are computed as

$$\tilde{r}_{\mathrm{lpc},i}(n) = \frac{1}{N_{\mathrm{block}} - i - 1} \sum_{k=0}^{N_{\mathrm{block}}-i-1} s(n+k)s(n+k+i), \tag{5.13}$$

where $i = [0, \ldots, N_{\mathrm{lpc}} - 1]$, N_{block} is the length of the extracted block and $s(n)$ is the speech signal at time index n. For the respective orders we have the constraint

$$N_{\mathrm{block}} \geq N_{\mathrm{lpc}}. \tag{5.14}$$

Then the normalized autocorrelation coefficients of the narrowband input signal are

$$\mathbf{r}_{\mathrm{lpc}}(n) = \frac{\tilde{\mathbf{r}}_{\mathrm{lpc}}(n)}{\tilde{r}_{\mathrm{lpc},0}(n)}. \tag{5.15}$$

From these coefficients we can now calculate predictor coefficients by applying (3.30), cepstral coefficients by first computing the LPC coefficients and then applying (3.84) or LSFs by following the instructions in Sect. 3.2.5. This modification has the effect that more focus in terms of the respective optimization criterion is placed on the narrowband spectral envelope within the passband between approx. 300–3,400 Hz since all envelopes are more or less equal to the perturbance outside the passband. The additive perturbance for the autocorrelation coefficients can alternatively be computed in the time domain by computing the autocorrelation corresponding to (5.13). This results in the identical perturbance for $N_{\mathrm{DFT}} \geq N_{\mathrm{FIR}}$ and $N_{\mathrm{lpc}} \leq N_{\mathrm{DFT}} - N_{\mathrm{FIR}}$. Figure 5.6 shows the result of applying such a modification. In Chap. 6 the benefit of this additive perturbance can be seen if comparing the average distance of the method with and without the application of the perturbance. Since the broadband codebook is equal in both cases, the classification has obviously improved if the average distance has dropped.

The increased robustness can be seen in the following example. Figure 5.7 shows the spectrograms of two telephone bandlimited speech signals which have been upsampled from 8 kHz to 11,025 Hz by a low quality sample rate converter, as depicted in the upper graphic, and by a high quality sample rate converter, as depicted in the lower graphic. The imaging components in the upper plot of Fig. 5.7 can clearly be seen above 3,400 Hz. These imaging components cause the spectral envelopes of a short-term spectral analysis of

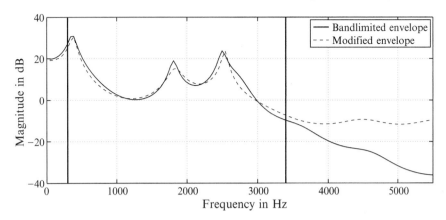

Fig. 5.6. Example of a narrowband spectral envelope and its modified pendant. Note that the envelopes are almost identical within the passband which is indicated by the two *vertical lines*

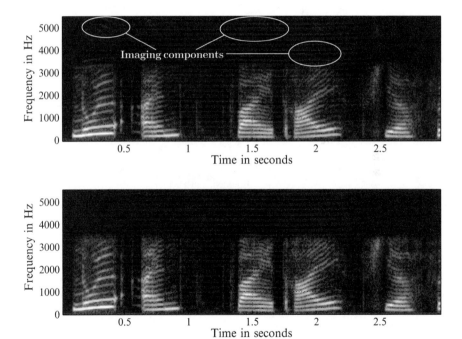

Fig. 5.7. Spectrogram of a telephone bandlimited speech signal being upsampled from 8 kHz to 11,025 Hz by a low quality sample rate converter (*upper plot*) and a high quality sample rate converter (*lower plot*)

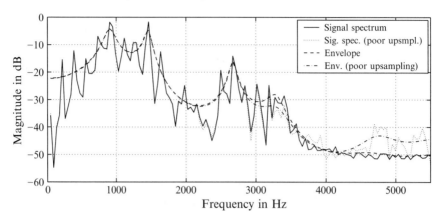

Fig. 5.8. Short-term spectrum of a frame of the speech signals depicted in Fig. 5.7 and their respective spectral envelopes

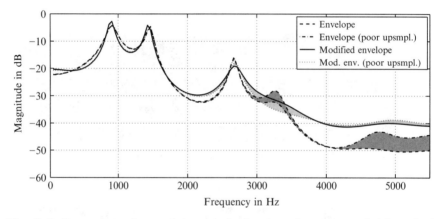

Fig. 5.9. Spectral envelopes of the signals that have been upsampled by a high quality and a low quality sample rate converter and the respective modified spectral envelopes

the two signals to show significant differences as can be seen in Fig. 5.8. If we now consider a codebook search for the envelope fitting the input envelope represented by the envelope of the speech signal that has undergone a low quality sample rate conversion best, the advantage of the autocorrelation manipulation becomes obvious. The distance between the two envelopes can be represented by the area between the two curves that have to be compared. This is depicted in Fig. 5.9. Recapitulatory we can state that applying the perturbance leads to lower sensitivity to low quality sample rate converters or differences in telephone bandpasses during training and operation and therefore increases the robustness of the classification.

5.3 Estimation of the Broadband Spectral Envelope Using Codebooks

After extracting the narrowband spectral envelope there exist a couple of methods for estimating the broadband spectral envelope from this information. One possible method which we will describe now in more detail is the use of a codebook [Yoshida 94], [Unno 05]. To be more specific the codebook consists of two combined codebooks, a narrowband codebook and a broadband codebook which are trained jointly [Epps 98]. By determining the most likely narrowband envelope within the codebook the broadband envelope is automatically obtained. Figure 5.10 illustrates this process. In this special case illustrated a weighting of the broadband spectral envelopes proportional to the inverse distance of the input envelope to the narrowband entries has been performed using LSFs (see Sect. 5.3.4) (compare [Cheng 94]). For the training of the codebooks an algorithm based on the LBG-algorithm (see Appendix B) has been applied [Linde 80, Cuperman 85]. Joint training of the narrowband and broadband codebooks in this context means that the real training of the centroids was done using the broadband data. The corresponding narrowband vectors of the broadband data that have been classified to one broadband centroid are then used for computing the corresponding narrowband centroid. The training can alternatively be based on the narrowband coefficients [Jax 02a]. The reason for the training based on the broadband coefficients used within this work is the fact that the broadband centroids would become blurred due to suboptimal quantization in the narrowband training data set and would therefore lead to inaccurate broadband spectral envelopes for the extension. If the codebook training is based on the broadband coefficients however, the inaccuracy occurs in the narrowband centroids and therefore in the classification.

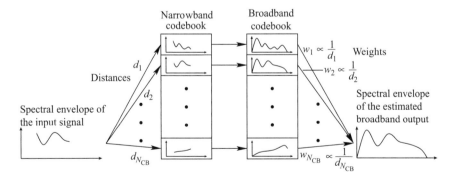

Fig. 5.10. Usage of two jointly trained codebooks for broadband spectral envelope estimation

Several codebooks using different kinds of features, e.g. AR coefficients, cepstral coefficients, MFCCs and LSFs, have been trained and will briefly be discussed in the following sections.

5.3.1 Codebook Training and Operation Using AR Coefficients

After the preparation of the training data set as described in Sect. 5.1 it is reasonable for some subsequent processing steps to analyze the statistics of the features and to normalize the features. This normalization has been performed in this book by using the following rule:

$$a_{i,\text{norm}} = \frac{a_i - m_{a_i}}{3\sigma_{a_1}}, \tag{5.16}$$

where

$$m_{a_i} = \frac{1}{N_{\text{feat}}} \sum_{k=0}^{N_{\text{feat}}-1} a_{i,k} \approx \mathcal{E}\left\{a_i\right\} \tag{5.17}$$

means computing the mean value (approximately the expectation value $\mathcal{E}\left\{\right\}$) of each coefficient a_i with N_{feat} being the amount of feature vectors that are included in the training data set and

$$\sqrt{\mathcal{E}\left\{(a_1 - \mathcal{E}\left\{a_1\right\})^2\right\}} \approx \sqrt{\frac{1}{N_{\text{feat}} - 1} \sum_{k=0}^{N_{\text{feat}}-1} (a_{1,k} - m_{a_1})^2} = \sigma_{a_1} \tag{5.18}$$

being the estimated standard deviation of the AR coefficient with index 1. This means that all AR-coefficients have been scaled by the standard deviation of the AR-coefficient with index 1. This might not be evident at first glance, but can be explained easily. Figure 5.11 shows the estimated standard deviation of each AR-coefficient for the used training data set consisting

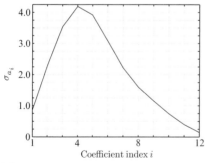

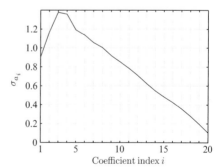

(a) Distribution of σ_{a_i} for narrowband AR-coefficients.

(b) Distribution of σ_{a_i} for broadband AR-coefficients.

Fig. 5.11. Distribution of σ_{a_i} for narrowband and broadband AR-coefficients

of $N_{\mathrm{feat}} = 4{,}794{,}592$ feature vectors or blocks respectively. The intuitive approach would be to scale each coefficient by its individual standard deviation. This would result in a flat distribution of σ_{a_i}. If we now apply for example a quadratic cost function within the LBG-algorithm for determining the distance of two vectors, the component for $i = 12$ would contribute as much as the component for $i = 3$ to the overall distance. Since the important information on the rough shape of the spectral envelope is contained in the first few coefficients it makes sense to stress the importance of theses coefficients by applying one common scaling factor to all coefficients.

Another processing step of the training data set comes up when not simulating the transmission line but recording after real transmission. Since effects like drop outs or comfort noise injection destroy the dependency between the broadband and narrowband features, the corresponding narrowband and broadband spectral envelopes have to be checked for similarity within the telephone band. This can be performed for example by the application of a simple quadratic cost function for computing the distance between two spectral envelopes within the telephone band. By deleting the feature-vector pairs that exceed a predefined distance threshold a new training data set with guaranteed correlation between broadband and narrowband counterparts is generated.

After the preparation of an appropriate training data set the codebook training can be initiated. The algorithm used for this codebook training is based on the LBG-algorithm described in Appendix B. Some minor modifications have been introduced to the basic algorithm. These modifications consist of the change of the perturbance for example. In the specific implementation no perturbance in the natural sense has been applied but the two nearest neighbors to a centroid have been searched for performing the initial splitting. As a distance measure based on the likelihood-ratio distance measure as introduced in Sect. 3.4.4 has been used

$$\tilde{d}_{\mathrm{LR}}(n, i) = \hat{\mathbf{a}}_i^T \mathbf{R}_{ss}(n) \hat{\mathbf{a}}_i. \tag{5.19}$$

This is not a real distance measure but it is sufficient during training and operation to quantize the feature vectors and to search for the entry producing minimum distance. The LPC order used for the broadband codebook has been $N_{\mathrm{lpc,bb}} = 20$ and $N_{\mathrm{lpc,nb}} = 12$ for the narrowband codebook respectively. Codebooks have been trained up to an amount of entries of $N_{\mathrm{cb}} = 4096$.

5.3.2 Codebook Training and Operation Using Cepstral Coefficients

For the training of another codebook cepstral coefficients have been used to represent the spectral envelope. For the narrowband codebook $N_{\mathrm{cc,nb}} = 18$ coefficients have been used and for the broadband codebook $N_{\mathrm{cc,bb}} = 30$ coefficients have been employed (according to the conventional amount of coefficients derived from LPC-coefficients $N_{\mathrm{ceps}} = \frac{3}{2} N_{\mathrm{lpc}}$). These coefficients have

been derived using LPC coefficients and applying the method described in
Sect. 3.2.3 for the computation of cepstral coefficients using LPC coefficients.
Additionally two codebook versions have been trained containing three more
features extra in the one case and a split codebook for voiced and unvoiced utterances in the other case. For representing the information on pitch frequency
the method described in Sect. 3.3.3 resulting in the ACI value as described in
(3.105) has been used. The value of the maximum of this modified, normalized autocorrelation, ACM (see (3.106)) has been used as well, representing
a speech feature correlating with the voicedness of the utterance (compare
[Epps 99, Jax 04b, Kornagel 02]) and also serving as confidence measure for
the pitch estimation. The third additional feature is the ratio of the power
within the higher frequency range to the power of the lower frequency range
(see Sect. 3.3.6). This measure has been exploited for the second type of codebook, differentiating voiced and unvoiced utterances, as well. The measure
can be computed employing (3.111) and the following relation:

$$W(m) = \sum_{n=m}^{m+N-1} s^2(n) = r_{ss}(0) = \frac{1}{N_{\text{DFT}}} \sum_{k=0}^{N_{\text{DFT}}-1} \left| S\left(e^{j\Omega_k}\right) \right|^2. \qquad (5.20)$$

By setting the summation limits to the appropriate values

$$W_{\text{low}}(m) = \frac{1}{k_{\text{low2}} - k_{\text{low1}} + 1} \sum_{k=k_{\text{low1}}}^{k_{\text{low2}}} \left| S\left(e^{j\Omega_k}\right) \right|^2, \qquad (5.21)$$

respectively

$$W_{\text{high}}(m) = \frac{1}{k_{\text{high2}} - k_{\text{high1}} + 1} \sum_{k=k_{\text{high1}}}^{k_{\text{high2}}} \left| S\left(e^{j\Omega_k}\right) \right|^2, \qquad (5.22)$$

with

$$0 \le k_{\text{low1}} < k_{\text{low2}} < k_{\text{high1}} < k_{\text{high2}} \le \frac{N_{\text{DFT}}}{2}, \qquad (5.23)$$

we can formulate the final ratio of the energy of the higher frequencies to the
lower ones

$$r_{\text{hlr}}(m) = \frac{W_{\text{high}}(m)}{W_{\text{low}}(m)}. \qquad (5.24)$$

This ratio also contains information on the voicedness of the actual utterance
but depends strongly on the limits chosen or the cut off frequencies of the
transmission line. Other features have also been investigated (see Sect. 3.3)
but after training a codebook containing the three most promising features
among those presented in Sect. 3.3 without according success, the application
of other features has been discarded (see Sect. 6.2).

All features have been normalized to zero mean and divided by 3σ according to the above mentioned subsequent processing (see Sect. 5.3.1). Additionally the maximum value of the autocorrelation function ACM has been

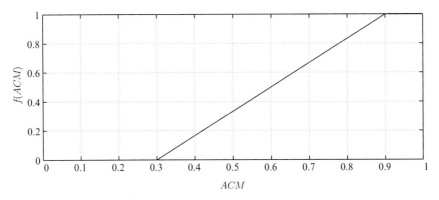

Fig. 5.12. Non-linear weighting function for the autocorrelation maximum ACM

weighted by a non-linear function (see Fig. 5.12). This weighting function forces the ACM value into almost only two states. Either to be equal to zero which means unvoiced or unreliable pitch estimation or to be equal to one meaning the pitch estimation is very reliable and the utterance is voiced. An example for the distribution of the ACM value before this weighting can be found in Appendix E. An enhanced cost function including the additional features for the training process as well as for the later operation has been developed. For the implementation without additional features the cepstral distortion measure as introduced in Sect. 3.4.3 has been used. The modified cost function using additional features takes the pitch frequency into account by using the index of the maximum of the autocorrelation function ACI. Furthermore the (quadratic) distance of the actual pitch to the mean pitch of each codebook entry is weighted by the non-linear weighted value of the maximum of the autocorrelation function. This is done to only take voiced utterances into account as the method used produces pitch values even in unvoiced segments

$$d_{\ell,\text{ceps,feat}}\left(\mathbf{c}(n), \mathbf{c}_{\ell,\text{cb}}\right) = \sum_{i=1}^{N_{\text{narrow}}} \left(c_i(n) - c_{i,\ell,\text{cb}}\right)^2$$
$$+(ACI_{\ell,\text{cb}} - ACI(n))^2 \cdot f(ACM(n))$$
$$+(ACM_{\ell,\text{cb}} - ACM(n))^2$$
$$+(r_{\text{hlr},\ell,\text{cb}} - r_{\text{hlr}}(n))^2, \tag{5.25}$$

for $\ell \in \{0, \dots, N_{\text{cb}} - 1\}$. In Appendix E the distribution of the above presented features as well as the distribution of the cepstral coefficients over the entire training data set is presented. As explained in the appendix the feature representing the pitch ACI for example is able to separate male and female speakers. Also the feature ACM in combination with r_{hlr} is able to classify into voiced and unvoiced utterances. Unfortunately this does not improve the quality of the estimated broadband spectral envelope as discussed in more detail in Sect. 5.7 and Chap. 6.

As already mentioned in Sect. 5.2 malclassifications lead to noticeable and bothersome artifacts. Another method for increasing the robustness against malclassifications can be applied during the operation mode using the statistics determined after the training phase. This method to reduce malclassifications comprises the observance of prior classifications and the exploitation of the conditional probability for a certain classification under the condition of the observation of a preceding classification. This probability can be estimated after having trained the codebook using the training corpus for determining the statistics. If we define the classification (giving the index of the codebook entry that produces the minimum distance to the current input feature vector) $\mathfrak{C}_{N_{\mathrm{cb}}}(\mathbf{c}(n))$ within the codebook of size N_{cb} of the actual input feature vector $\mathbf{c}(n)$ as

$$\mathfrak{C}_{N_{\mathrm{cb}}}(\mathbf{c}(n)) = \arg \min_{i=0}^{N_{\mathrm{cb}}-1} \left\{ d_{i,\mathrm{ceps}}(\mathbf{c}(n), \mathbf{c}_{i,\mathrm{cb}}) \right\}, \tag{5.26}$$

and by defining a counting function

$$\delta_{\ell,i}\left(\mathfrak{C}_{N_{\mathrm{cb}}}(\mathbf{c}(n-1)), \mathfrak{C}_{N_{\mathrm{cb}}}(\mathbf{c}(n))\right) = \begin{cases} 1, & \text{for } \mathfrak{C}_{N_{\mathrm{cb}}}(\mathbf{c}(n-1)) = \ell \\ & \text{and } \mathfrak{C}_{N_{\mathrm{cb}}}(\mathbf{c}(n)) = i, \\ 0, & \text{otherwise}, \end{cases} \tag{5.27}$$

we can build a transition matrix containing the elements

$$\sigma_{\ell,i} = \sum_{n=1}^{N_{\mathrm{feat}}-1} \delta_{\ell,i}\left(\mathfrak{C}_{N_{\mathrm{cb}}}(\mathbf{c}(n-1)), \mathfrak{C}_{N_{\mathrm{cb}}}(\mathbf{c}(n))\right), \tag{5.28}$$

for $\ell \in \{0, \ldots, N_{\mathrm{cb}}-1\}$ and $i \in \{0, \ldots, N_{\mathrm{cb}}-1\}$. Now we can determine the transition probabilities by using

$$p\left(\mathfrak{C}_{N_{\mathrm{cb}}}(\mathbf{c}(n)) = i \,\middle|\, \mathfrak{C}_{N_{\mathrm{cb}}}(\mathbf{c}(n-1)) = \ell\right)$$

$$= \frac{p\left(\mathfrak{C}_{N_{\mathrm{cb}}}(\mathbf{c}(n)) = i, \, \mathfrak{C}_{N_{\mathrm{cb}}}(\mathbf{c}(n-1)) = \ell\right)}{p\left(\mathfrak{C}_{N_{\mathrm{cb}}}(\mathbf{c}(n-1)) = \ell\right)}$$

$$\approx \frac{\sigma_{\ell,i} \left/ \sum_{\ell=0}^{N_{\mathrm{cb}}-1} \sum_{i=0}^{N_{\mathrm{cb}}-1} \sigma_{\ell,i}\right.}{\sum_{i=0}^{N_{\mathrm{cb}}-1} \sigma_{\ell,i} \left/ \sum_{\ell=0}^{N_{\mathrm{cb}}-1} \sum_{i=0}^{N_{\mathrm{cb}}-1} \sigma_{\ell,i}\right.} \tag{5.29}$$

$$= \frac{\sigma_{\ell,i}}{\sum_{i=0}^{N_{\mathrm{cb}}-1} \sigma_{\ell,i}} \tag{5.30}$$

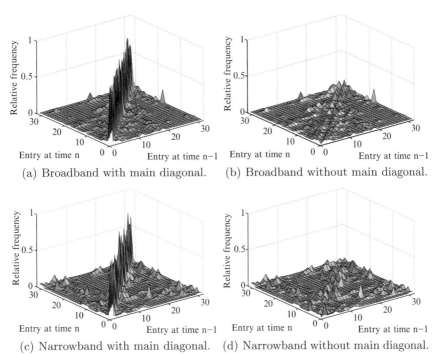

(a) Broadband with main diagonal.   (b) Broadband without main diagonal.

(c) Narrowband with main diagonal.   (d) Narrowband without main diagonal.

Fig. 5.13. Relative transition frequency for a codebook containing 32 entries of (**a**) the codebook entries for using the broadband codebook and broadband data for classification and (**c**) the codebook entries for using the narrowband codebook and narrowband classification. For the figures (**b**) and (**d**) the main diagonal has been set to zero for reasons of clarity

$$= \hat{p}\Big(\mathfrak{C}_{N_{cb}}\big(\mathbf{c}(n)\big) = i \,\Big|\, \mathfrak{C}_{N_{cb}}\big(\mathbf{c}(n-1)\big) = \ell\Big).$$

Figure 5.13 shows the transition probabilities using a virtual classification on the broadband codebook and the real classification using the narrowband codebook for $N_{cb} = 32$, respectively. The difference between those two transition patterns is due to malclassifications and can be reduced by applying the transition probabilities, derived by the virtual classification using the broadband codebook, within a modified distance measure. Several approaches have been tested, e.g. the following distance measure

$$d_{i,\mathrm{ceps}}\big(\mathbf{c}(n), \mathbf{c}(n-1), \mathbf{c}_{i,\mathrm{cb}}\big) =$$

$$\left(1 - \hat{p}\Big(\mathfrak{C}_{N_{cb}}\big(\mathbf{c}(n)\big) = i \,\Big|\, \mathfrak{C}_{N_{cb}}\big(\mathbf{c}(n-1)\big) = \ell\Big)\right) \cdot \sum_{k=1}^{N_{\mathrm{ceps}}} \big(c_k(n) - c_{k,i,\mathrm{cb}}\big)^2,$$

$$(5.31)$$

for $i \in \{0, \ldots, N_{cb} - 1\}$. Experiments using this transition probability showed no significant improvement of the speech quality and have therefore been abandoned within this work.

5.3.3 Codebook Training and Operation Using MFCCs

Since MFCCs are the most common choice in speech recognition applications, these coefficients have also been studied. Different from the above described coefficients and the following LSFs these coefficients have not been calculated on the base of autocorrelation coefficients but using the method described in Sect. 3.2.4. For the broadband codebook $N_{mfcc,bb} = 28$ coefficients and $N_{mfcc,nb} = 13$ coefficients for the narrowband codebook have been used, respectively. The feature vectors have been normalized as described above in (5.16) and (5.18). As a cost function for the training of a codebook using MFCCs the cepstral distance

$$d_{mfcc}\left(\mathbf{c}_{mfcc}, \hat{\mathbf{c}}_{mfcc}\right) = \sum_{k=1}^{N_{mfcc}} \left(c_{k,mfcc} - \hat{c}_{k,mfcc}\right)^2, \tag{5.32}$$

as introduced in Sect. 3.4.3 has been used. As discussed in more detail in Sect. 5.7 and Chap. 6 the performance using MFCCs is not superior to using cepstral coefficients. The fact that the computational effort for computing MFCCs exceeds the effort needed for computing cepstral coefficients makes the MFCCs unsuitable for the BWE application.

5.3.4 Codebook Training and Operation Using LSFs

The application of LSFs is inspired by the broad use within the field of speech coding and the properties such as robustness when quantizing feature vectors meaning that always a stable all-pole filter is generated since we are only manipulating the angles of zeros that lie on the unit circle. The LSF coefficients are calculated using LPC coefficients and the method presented in Sect. 3.2.5. For finding the roots of the polynomials the method described in Appendix D has been used. As a distance measure during training and operation the cost function

$$d_m\left(\boldsymbol{\alpha}_{in}, \boldsymbol{\alpha}_{m,cb}\right) = \sum_{i=0}^{N_{lsf,nb}-1} \left|\alpha_{i,in}\left(n\right) - \alpha_{i,m,cb}\right|, \text{ for } m \in \{0, \ldots, N_{cb} - 1\}, \tag{5.33}$$

according to (3.118) for $p = 1$ has been employed. For representing the broadband spectral envelopes $N_{lsf,bb} = 20$ coefficients and for representing the narrowband spectral envelopes $N_{lsf,nb} = 12$ coefficients have been used, respectively. The feature vectors have been normalized according to the above discussed normalization. For the operation mode the output LSFs have been computed as a sum over all codebook entries

$$\alpha_{i,\text{out}} = \sum_{m=0}^{N_{\text{cb}}-1} w_m \cdot \alpha_{i,m,\text{bcb}} \text{ , for } i = 0, \ldots, N_{\text{lsf,bb}} - 1, \tag{5.34}$$

weighted by an expression that is proportional to the inverse of the respective distance the narrowband codebook entry produced

$$w_m = \frac{e^{-\gamma d_m}}{\sum_{i=0}^{N_{\text{cb}}-1} e^{-\gamma d_i}} \text{ , for } m = 0, \ldots, N_{\text{cb}} - 1, \tag{5.35}$$

where γ is a parameter for steering the steepness of the exponential function and N_{cb} being the amount of codebook entries. This method is known as codebook interpolation and can be found in various flavors, e.g. in [Hu 05].

Since for calculating the zeros of the two polynomials (see Sect. 3.2.5 and Appendix D) some computational effort has to be spent and the LSFs didn't outperform the other representations of the spectral envelope the use of this set of coefficients has also been discarded.

5.4 Estimation of the Broadband Spectral Envelope Using Neural Networks

A completely different approach from the application of a codebook pair is the use of artificial neural networks [Zaykovskiy 05, Parveen 04, Shahina 06, Uncini 99]. Artificial neural networks are a very popular choice in the field of forecast, control tasks, and pattern recognition, which are in fact similar to the task of extrapolating a broadband spectral envelope out of a bandlimited one [Rojas 96, Nauck 96, StatSoft 02]. The idea is to train an artificial neural network offline by providing pairs of bandlimited and broadband spectral envelopes or their parametric representations, respectively. The artificial neural networks used in this book comprise feed forward networks (multilayer perceptrons or MLPs) consisting of one input layer, one hidden layer, and one output layer. Figure 5.14 shows such a feed forward network consisting of one hidden layer in this case.

Generally artificial neural networks can comprise $\nu \in \{0, \ldots, N_{\text{layer}} - 1\}$ layers. One input and one output layer and $N_{\text{layer}} - 2$ hidden layers. Within the input layer the nodes do not perform any processing. The output $a_{\lambda,0}$ of the input neurons equals the input. Processing is only done within the nodes of the hidden layer(s) and the output layer. The amount of nodes $\lambda \in \{0, \ldots, N_\nu - 1\}$ within one layer ν is independent of the amount of nodes within the preceding or succeeding layer. The amount of inputs and outputs equals the amount of coefficients used for the narrowband spectral envelope representation and broadband one, respectively. The function that is implemented in each node λ within the hidden layer or within the output layer ν producing the output is

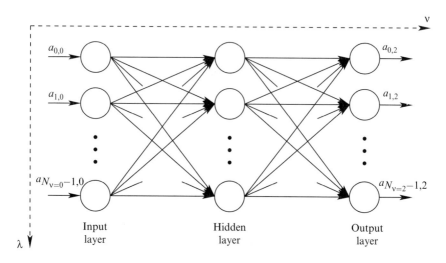

Fig. 5.14. Structure of a feed forward network with one hidden layer

$$a_{\lambda,\nu} = f_{\text{act}}\left(u_{\lambda,\nu}\right) = f_{\text{act}}\left(\theta_{\lambda,\nu} + \sum_{\mu=0}^{N_{\nu-1}-1} a_{\mu,\nu-1}\, w_{\mu,\nu-1,\lambda}\right), \tag{5.36}$$

$$\text{for } \nu \in \{1,\dots,N_{\text{layer}}-1\},\ \lambda \in \{0,\dots,N_\nu-1\},$$

where $a_{\lambda,\nu}$ denotes the output of the node with index λ in layer ν. With N_ν representing the amount of nodes in layer ν. The quantity $w_{\mu,\nu-1,\lambda}$ denotes the weight which is applied to the output of node μ of the preceding layer $\nu - 1$ which serves then after weighting as input of node λ of the actual layer in process ν. The parameter $\theta_{\lambda,\nu}$ denotes the bias of each node. As an activation function a tanh function as well as an exponential function, both are sigmoid functions, have been used

$$f_{\text{act}}(x) = \tanh(x), \tag{5.37}$$

$$f_{\text{act}}(x) = \frac{1}{1 + e^{-x}}. \tag{5.38}$$

Fig. 5.15 shows the function of a single node within the hidden layer(s). The characteristics of the sigmoid functions that have been used as activation functions can be seen in Fig. 5.16. All in all they are very similar except for the different normalization of the input data needed by the two functions.

5.4.1 Training and Operation Using Cepstral Coefficients

The training of the neural networks is accomplished similar to the training of the codebooks by providing pairs of bandlimited and broadband data. The

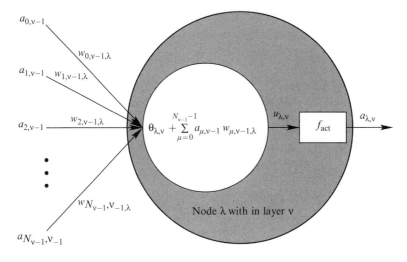

Fig. 5.15. Function of a neuron

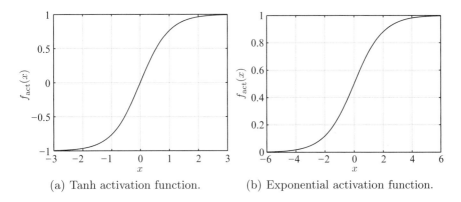

(a) Tanh activation function.   (b) Exponential activation function.

Fig. 5.16. Activation functions

algorithm used for training the neural network is described in more detail in Appendix C. The training itself then has to be performed very carefully to avoid overtraining. Overtraining denotes the optimization on the training set only without further generalization. This can be observed by using a validation data set to control if the network still is generalizing its task or beginning to learn the training data set by heart. Figure 5.17 shows such a typical behavior. The optimum iteration to stop the training is marked in Fig. 5.17. This optimum is characterized by the minimum overall distance between the actual and the desired output of the artificial neural network produced using the validation data set as input. For the approach using a neural network cepstral coefficients have been used. This is due to the optimization criterion of the algorithm used to train the weights of the network. Since the optimization

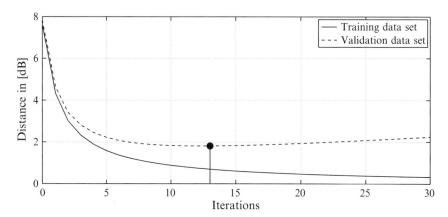

Fig. 5.17. Typical progression of the distance between actual output and desired output of the artificial neural network during the training phase

criterion (see (C.11)) equals the cepstral distance measure

$$d_{\mathrm{ceps}}\left(\mathbf{c}, \hat{\mathbf{c}}\right) = \sum_{k=1}^{N_{\mathrm{ceps}}} \left(c_k - \hat{c}_k\right)^2, \tag{5.39}$$

it is reasonable to use cepstral coefficients for training and operating a neural network.

Another crucial part is the proper normalization of the training data set. The result after training the neural network is very sensitive to the normalization used.

When it comes to the operation mode of the artificial neural network, after the training phase has been completed, there are two major characteristics of such a network that have to be mentioned:

- The first one is the low computational complexity needed by such an artificial neural network if not too many layers or nodes are used respectively. The network used in this work comprises three layers. The input layer consists of 17 nodes, the hidden and the output layer consist of 29 nodes each. The activation function according to (5.38) has been applied. Such networks are able to learn complex tasks using comparatively few layers and neurons. This is an advantage over codebook approaches since the computational effort used to evaluate a distance measure for each codebook entry is omitted using artificial neural networks.

- On the other hand artificial neural networks do not offer possibilities to interfere manually the way codebook approaches do. For example by observing the distance between the narrowband spectral envelope of the input signal and the narrowband codebook entry producing minimum distance one is able to predict if the bandwidth extension runs out of the rudder completely and therefore switch off such a system. Since there is no such thing as a reference using neural networks this is not possible.

Another property is that the output all-pole filter that is described by the cepstral coefficients is not guaranteed to be stable as it would be the case using a codebook.

5.5 Estimation of the Broadband Spectral Envelope Using Linear Mapping

The approaches presented for the estimation of the broadband spectral envelope up to now comprise only non-linear processing. A very simple and intuitive approach is the search for a linear dependency of the bandlimited spectral envelope of the input signal and the broadband counterpart. Such approaches are known as *linear mapping* [Jiménez-Fernàndez 98, Zaykovskiy 04]. In the following we will briefly describe the derivation of such a dependency and develop an extension of this approach in the following sections.

If the parameter set containing the bandlimited envelope information is described by a vector

$$\mathbf{x}(n) = \left[x_0(n),\, x_1(n),\, \ldots,\, x_{N_x-1}(n)\right]^T \tag{5.40}$$

and its wideband counterpart by a vector

$$\mathbf{y}(n) = \left[y_0(n),\, y_1(n),\, \ldots,\, y_{N_y-1}(n)\right]^T, \tag{5.41}$$

then a linear estimation scheme can be realized by a simple linear operation

$$\hat{\mathbf{y}}(n) = \mathbf{W}\left(\mathbf{x}(n) - \mathbf{m}_x\right) + \mathbf{m}_y. \tag{5.42}$$

The entries of the vectors $\mathbf{x}(n)$ and $\mathbf{y}(n)$ could be predictor coefficients, cepstral coefficients, or line spectral frequencies. In this book cepstral coefficients have been used. The multiplication of the bandlimited feature vector $\mathbf{x}(n)$ with the $N_y \times N_x$ matrix $\mathbf{W}$ can be interpreted as a set of N_x FIR filter operations (compare [Avendano 95] and [Bansal 05]). Each row of $\mathbf{W}$ corresponds to an impulse response which is convolved with the signal vector $\mathbf{x}(n)$ resulting in one element of the estimated wideband feature vector $\hat{\mathbf{y}}(n)$. As common in linear estimation theory the mean values of the feature vectors $\mathbf{m}_x$ and $\mathbf{m}_y$ are estimated within a preprocessing stage.

5.5.1 Training and Operation Using Cepstral Coefficients

For obtaining the matrix $\mathbf{W}$ first a cost function has to be specified. A very simple approach would be the minimization of the sum of the squared errors over a large database:

$$F(\mathbf{W}) = \sum_{n=0}^{N-1} ||\mathbf{y}(n) - \hat{\mathbf{y}}(n)||^2 \to \min . \tag{5.43}$$

If we define the entire data base consisting of N zero-mean feature vectors by two matrices

$$\mathbf{X} = \left[\mathbf{x}(0) - \mathbf{m}_x, \mathbf{x}(1) - \mathbf{m}_x, \ldots, \mathbf{x}(N-1) - \mathbf{m}_x \right], \tag{5.44}$$

$$\mathbf{Y} = \left[\mathbf{y}(0) - \mathbf{m}_y, \mathbf{y}(1) - \mathbf{m}_y, \ldots, \mathbf{y}(N-1) - \mathbf{m}_y \right], \tag{5.45}$$

the optimal solution is given by (see [Luenberger 69])

$$\mathbf{W}_{\text{opt}} = \mathbf{Y}\,\mathbf{X}^T \left(\mathbf{X}\,\mathbf{X}^T \right)^{-1} . \tag{5.46}$$

Since the sum of the squared differences of cepstral coefficients is a well-accepted distance measure in speech processing often cepstral coefficients are utilized as feature vectors. Even if the assumption of the existence of a single matrix $\mathbf{W}$ which transforms all kinds of bandlimited spectral envelopes into their broadband counterpart is quite unrealistic, this simple approach results in astonishing good results.

However, the basic single matrix scheme can be enhanced by using several matrices, where each matrix was optimized for a certain type of feature class [Nakatoh 97]. In a two matrices scenario one matrix $\mathbf{W}_v$ can be optimized for voiced sounds and the other matrix $\mathbf{W}_u$ for non-voiced sounds for example. In this case it is first checked to which class the current feature vector $\mathbf{x}(n)$ belongs. In a second stage the corresponding matrix is applied to generate the estimated wideband feature vector[1]

$$\hat{\mathbf{y}}(n) = \begin{cases} \mathbf{W}_v \left(\mathbf{x}(n) - \mathbf{m}_x \right) + \mathbf{m}_y , & \text{if } \mathbf{x}(n) \text{ is classified as voiced,} \\[2mm] \mathbf{W}_u \left(\mathbf{x}(n) - \mathbf{m}_x \right) + \mathbf{m}_y , & \text{else.} \end{cases} \tag{5.47}$$

The classification of the type of sound can be performed by analyses such as zero-crossing rate [Deller Jr. 00] or gradient index [Jax 02a] or any other of the scalar speech parameters derived in Sect. 3.3 (see [Chennoukh 01]).

5.6 Combined Approach Using Codebook Classification and Linear Mapping

Prosecuting consequently the idea presented above of classifying the speech signal first and then providing specific matrices leads to the awareness that linear mapping can be applied as a postprocessing stage of codebook approaches. In this case the nonlinear mapping between bandlimited and wideband feature

---

[1] Besides two different matrices $\mathbf{W}_v$ and $\mathbf{W}_u$ also different mean vectors $\mathbf{m}_y$ and $\mathbf{m}_x$ can be applied for voiced and unvoiced frames.

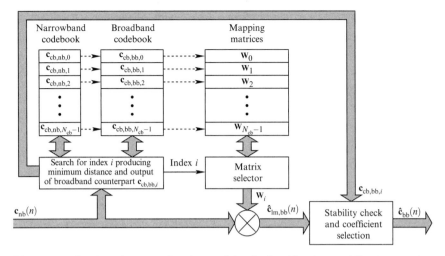

Fig. 5.18. Combined approach using codebook classification and linear mapping

vectors is modeled as a piecewise linear mapping. Figure 5.18 shows such a design. First the narrowband codebook is searched for the narrowband entry or its index i, respectively, producing minimum distance to the actual input feature vector considering a specific distance measure (see Sect. 3.4)

$$i(n) = \mathfrak{C}_{N_{\mathrm{cb}}}\big(\mathbf{c}_{\mathrm{nb}}(n)\big) = \arg \min_{k=0}^{N_{\mathrm{cb}}-1} \big\{ d_{\mathrm{ceps}}\big(\mathbf{c}_{\mathrm{nb}}(n), \mathbf{c}_{\mathrm{cb,nb},k}\big) \big\}. \tag{5.48}$$

Here the cepstral distance measure as defined in (3.123) has been used. The coefficient vectors are defined as[2]

$$\mathbf{c}_{\mathrm{nb}}(n) = \Big[c_{\mathrm{nb}}^{(0)}(n), c_{\mathrm{nb}}^{(1)}(n), \ldots, c_{\mathrm{nb}}^{(N_{\mathrm{ceps,nb}}-1)}(n) \Big]^{\mathrm{T}}, \tag{5.49}$$

$$\mathbf{c}_{\mathrm{cb,nb},i} = \Big[c_{\mathrm{cb,nb},i}^{(0)}, c_{\mathrm{cb,nb},i}^{(1)}, \ldots, c_{\mathrm{cb,nb},i}^{(N_{\mathrm{ceps,nb}}-1)} \Big]^{\mathrm{T}}, \tag{5.50}$$

$$\mathbf{c}_{\mathrm{cb,bb},i} = \Big[c_{\mathrm{cb,bb},i}^{(0)}, c_{\mathrm{cb,bb},i}^{(1)}, \ldots, c_{\mathrm{cb,bb},i}^{(N_{\mathrm{ceps,bb}}-1)} \Big]^{\mathrm{T}}, \tag{5.51}$$

$$\hat{\mathbf{c}}_{\mathrm{lm,bb}}(n) = \Big[\hat{c}_{\mathrm{lm,bb}}^{(0)}(n), \hat{c}_{\mathrm{lm,bb}}^{(1)}(n), \ldots, \hat{c}_{\mathrm{lm,bb}}^{(N_{\mathrm{ceps,bb}}-1)}(n) \Big]^{\mathrm{T}}, \tag{5.52}$$

$$\hat{\mathbf{c}}_{\mathrm{bb}}(n) = \Big[\hat{c}_{\mathrm{bb}}^{(0)}(n), \hat{c}_{\mathrm{bb}}^{(1)}(n), \ldots, \hat{c}_{\mathrm{bb}}^{(N_{\mathrm{ceps,bb}}-1)}(n) \Big]^{\mathrm{T}}, \tag{5.53}$$

for $i \in \{0, \ldots, N_{\mathrm{cb}} - 1\}$. With N_{cb} being the amount of codebook entries which is equal for the narrowband and broadband codebook as well as for the amount of mapping matrices used. $N_{\mathrm{ceps,nb}}$ and $N_{\mathrm{ceps,bb}}$ denote the order of the narrowband cepstral vectors and broadband ones, respectively.

---

[2] Note that the notation has changed slightly. The upper index now denotes the component index of the vector, whereas the lower index describes the entry index of the respective narrowband or broadband codebook.

Then an initial estimated broadband spectral envelope is generated by multiplying the matrix $\mathbf{W}_i$ corresponding to the codebook entry pair $[\mathbf{c}_{\mathrm{cb,nb},i},\ \mathbf{c}_{\mathrm{cb,bb},i}]$, with the cepstral representation of the input narrowband spectral envelope $\mathbf{c}_{\mathrm{nb}}(n)$ resulting in $\hat{\mathbf{c}}_{\mathrm{lm,bb}}(n)$. In the final block depicted on the right hand side of Fig. 5.18 the all-pole filter corresponding to $\hat{\mathbf{c}}_{\mathrm{lm,bb}}(n)$ is checked for stability (see Appendix D) and depending on the result either the result of the linear mapping operation or the corresponding entry of the broadband codebook $\mathbf{c}_{\mathrm{cb,bb},i}$ is output as $\hat{\mathbf{c}}_{\mathrm{bb}}(n)$

$$
\hat{\mathbf{c}}_{\mathrm{bb}}(n) = \begin{cases} \mathbf{W}_i\big(\mathbf{c}_{\mathrm{nb}}(n) - \mathbf{m}_{\mathrm{nb},i}\big) + \mathbf{m}_{\mathrm{bb},i}, & \text{if all poles of the resulting filter} \\ & \text{are inside the unit circle,} \\ \mathbf{c}_{\mathrm{cb,bb},i}, & \text{otherwise.} \end{cases}
$$

$$(5.54)$$

The advantage of a codebook in combination with linear mapping is the possibility of using an algorithm that is able to learn autonomously the most characteristic spectral envelopes within a predetermined training material. This means that the task of extrapolating the spectral envelope for each mapping matrix is very specific. Another advantage of such a system is the possibility of having a fall back plane represented by the presence of a codebook containing the broadband spectral envelopes in parametric representation. If the application of a mapping matrix produces a non-stable all-pole filter an algorithm for preventing this is able to replace this all-pole filter by the respective entry of the broadband codebook. This is also very efficient since only the stability of the resulting filter has to be checked and no sumptuous algorithm for stabilizing an unstable filter has to be employed. Since the entries of the broadband codebook are guaranteed to be stable the whole algorithm is guaranteed to deliver a stable all-pole filter. This approach is first published within this book. Another advantage is the computational load that can be saved by significantly reducing the size of the codebook, since this is possible without quality decrease in comparison to an approach using only codebooks. The amount of additional memory needed for the linear mapping matrices is $N_{\mathrm{cb}}N_{\mathrm{ceps,nb}}N_{\mathrm{ceps,bb}}$ words for a codebook containing N_{cb} entries.

5.6.1 Training and Operation Using Cepstral Coefficients

The training of this combined approach can be split into two separate training stages. The training of the codebook is independent of the succeeding linear mapping matrices and can be conducted as described in Sect. 5.3.2. Then the entire training data is grouped into N_{cb} sets containing all feature vectors classified to the specific codebook entries. Now for each subset of the entire training material a single mapping matrix is trained according to the method described in Sect. 5.5.1. The correspondence of the linear mapping matrices and the codebook entries is indicated by the dashed arrows in Fig. 5.18. In Fig. 5.19 the function principle of the combined approach using a preclassification by a codebook and afterwards doing an individual linear mapping corresponding to each codebook entry is illustrated. Each little dot in Fig. 5.19a

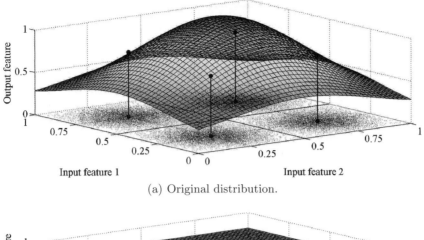

(a) Original distribution.

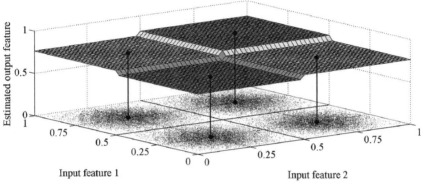

(b) Approximated distribution using a codebook approach.

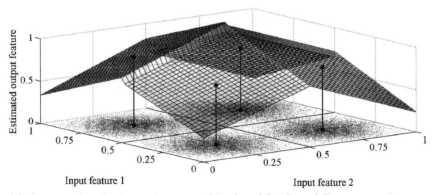

(c) Approx. distribution using a combined codebook and linear mapping approach.

Fig. 5.19. Illustration of the effect of a codebook approach (**b**) and a combined approach with additional linear mapping (**c**) for rebuilding a given distribution (**a**)

at the bottom represents a data point of the training material in the two dimensional input feature space (here for this example we limit ourselves to a two dimensional space). The lower big dot represents the centroid of each cluster which is the euclidean mean over all input vectors within this cluster (here already four clusters have emerged). The surface in Fig. 5.19a represents the real interrelation of the input vectors to one feature of the output vectors. So this would be the desired function for the extrapolation task. Since we do not know this function we map the output feature of all vectors that fall into one cell to the output feature of the centroid of this cell represented by the big black dot on the output feature surface (each line combines the centroid of the input feature clusters to the output feature vector corresponding to this centroid). This is true for the application of a codebook. To point out the effect of this the approximated output feature distribution for a codebook approach is depicted in Fig. 5.19b. We can see that this codebook approach makes no difference if the current input feature vector is placed near or far from the centroid of the cluster, but only uses the information that the input feature vector has been classified to this cluster. If we now combine a codebook classification with linear mapping as illustrated in Fig. 5.19c a plane is placed according to the data points within each cell with minimum overall distance to the original surface resulting in less error when processing an input vector which is close to a cell border.

5.7 Discussion

In this chapter we have provided an overview on the approaches used within this work for estimating the broadband spectral envelope of a bandlimited signal using the narrowband spectral envelope and some optional additional features. We have addressed codebook approaches, neural network approaches, approaches using linear mapping, as well as mixtures. A method for improving the classification robustness using an additive perturbance for the autocorrelation coefficients has been derived. The use of additional features to the parametric representation of the spectral envelope has been investigated as well as the use of different sets of parametric representations of the spectral envelope. Another approach for improved robustness against malclassifications has been introduced using the conditional probability of a certain classification under the observance of a certain classification in the preceding processing step. A detailed evaluation of the approaches will be given in Sect. 6.2. In advance we can state that, concerning the codebook approaches, the additional effort to increase the robustness against malclassifications using additional features or the observance of the preceding classification do not increase quality in a way that justifies the additional expense. The neural network approaches are able to perform in a satisfying manner. The major problem however is that neural networks do not offer the possibility to interfere as it is the case with codebook approaches or linear mapping. Furthermore the outcome of a network training is somewhat unpredictable. Depending on the amount of layers, neurons, the activation function used, and the normalization of the training data very divergent results occur. This is the reason why the neural network approach has not been pursued further although quite acceptable results have been achieved. The method that delivered the most promising results was the combination of a codebook classification with mapping matrices that have been trained class-specifically. This approach combines some convenient attributes. If the resulting synthesis filter is unstable we still have a broadband codebook entry for extending the envelope. The size of the codebook can be reduced significantly while maintaining a high quality in comparison to plain codebook approaches (see Sect. 6.2).

6

Quality Evaluation

This chapter deals with the evaluation of the quality produced by the different approaches for the tasks of extending the excitation signal and estimating the broadband spectral envelope. All in all we can differentiate between objective and subjective criteria. The objective quality criteria that have been used within this book have been introduced in Sect. 3.4. The subjective evaluation is done by subjects listening to the various approaches followed by a voting for the different approaches. This can be done in miscellaneous fashions. The specific method used for the subjective quality evaluation will be discussed within the respective sections. The examination of subjective quality evaluation will close with a significance analysis. At the end of the evaluation of each component for extending the bandwidth of a bandlimited speech signal the analysis of the rank correlation between the objective and subjective quality criteria will give information on the usability of the various objective quality criteria respectively the usability of objective quality criteria in general. A very popular speech quality measure (PESQ; Perceptual Evaluation of Speech Quality [ITU 01]) that tries to simulate a subjective listening test has been examined but has been discarded since it gave no reasonable results, which is due to the different kind of scenario it was designed for. Other examinations can be found in [Kallio 02].

6.1 Evaluation of the Excitation Signal Extension

Corresponding to the separation within the source-filter model into an excitation signal and the following coloration with a spectral envelope we will begin with analyzing the different methods introduced in Chap. 4 for the extension of the excitation signal [Iser 06]. We will present objective as well as subjective quality evaluation methods. The evaluation is done separately for the upper and the lower extension region. This is done due to the observation that the same method produces rather different satisfactory results in the sense of sound quality when applied to the lower or upper extension region, respectively. The following section deals with the objective quality evaluation.

6.1.1 Objective Quality Criteria

For the objective evaluation of the excitation signal the distance measures introduced in Sect. 3.4 have been used. For this task the broadband spectral envelope of the broadband signal has been calculated and used to whiten the narrowband signal as well as the broadband signal to extract the narrowband and broadband excitation signal, respectively. Then the different algorithms to extend the narrowband excitation signal have been applied. This extended excitation signal has then been compared in the sense of the different distance measures with the excitation signal of the original broadband signal. This has been done separately for the upper and for the lower extension region. For this purpose the interval over which the distance measure is evaluated has been modified. This results, regarding the log spectral distance as an example, in

$$d_{\mathrm{LSD},\Omega_1,\Omega_2}\left(E_{\mathrm{bb}},\hat{E}_{\mathrm{bb}}\right) = \frac{1}{\Omega_2 - \Omega_1}\int\limits_{\Omega_1}^{\Omega_2}\left|20\log_{10}\left|\frac{E_{\mathrm{bb}}(e^{j\Omega})}{\hat{E}_{\mathrm{bb}}(e^{j\Omega})}\right|\right|\mathrm{d}\Omega. \qquad (6.1)$$

For the implementation the integral has been carried over to a sum in the DFT domain. Applying this principle to the distance measures introduced in Sect. 3.4 and evaluating the different approaches using the whole training data base on a frame by frame basis with succeeding averaging over the amount of processed frames with $\Omega_1 \hat{=} 3400$ Hz and $\Omega_2 \hat{=} 5512.5$ Hz, corresponding to half of the sampling rate, results in Fig. 6.1. for the upper extension band. For all applied distance measures, as can be seen in Fig. 6.1, the adaptive quadratic characteristic achieves the best results. The spectral shift used within the extension of the upper region has been implemented as a fixed spectral shift. The spectral shifting approach does not perform well. This is true for all applied distance measures as well. Similar to the spectral shifting approach, the approach using a white noise generator for extending the excitation signal is judged poor by all of the applied distance measures. The half-way rectification however is judged almost as good as the adaptive quadratic characteristic. Note that the algorithms compared in this book for each region represent a preselection of promising methods, that have been chosen performing a primal informal listening test.

Applying the same procedure for evaluating the objective quality of a preselection of promising algorithms for the lower extension region ($\Omega_1 \hat{=} 0$ Hz and $\Omega_2 \hat{=} 300$ Hz) results in Fig. 6.2. Here we can observe that the adaptive quadratic characteristic again performs very well for all distance measures evaluated. However the other results diverge. Therefore no meaningful assertion can be made. However one difference should be mentioned. The cubic

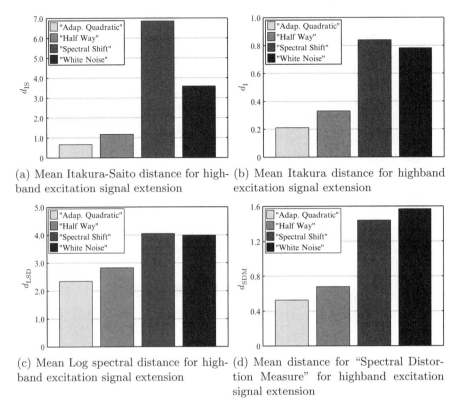

(a) Mean Itakura-Saito distance for high-band excitation signal extension

(b) Mean Itakura distance for highband excitation signal extension

(c) Mean Log spectral distance for high-band excitation signal extension

(d) Mean distance for "Spectral Distortion Measure" for highband excitation signal extension

Fig. 6.1. Results of the objective quality evaluation for the upper extension region for (**a**) Itakura–Saito, (**b**) Itakura, (**c**) Log spectral distortion and (**d**) "Spectral distortion Measure"

characteristic performs exceptionally poor concerning the "Spectral Distortion Measure". At first glance this might question the quality of the "Spectral Distortion Measure". But as we will see later on (see Sect. 6.1.3) this is not the case.

6.1.2 Subjective Quality Criteria

For evaluating the quality of different algorithms for extending the excitation signal in a subjective manner a listening test has been carried out [Pulakka 06]. Since it is not possible to judge the quality of a pure excitation signal the subjects had to listen to a regular speech signal. This speech signal was made up of the original signal within the telephone band and the original signal within the respective extension band, meaning when comparing the methods for extending the lower extension region the original signal reached from 300 up to 5,512.5 Hz and vice versa. The $M = 9$ subjects listened to $I = 20$ sets of

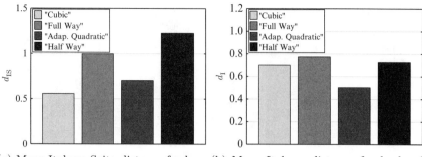

(a) Mean Itakura-Saito distance for low-band excitation signal extension

(b) Mean Itakura distance for lowband excitation signal extension

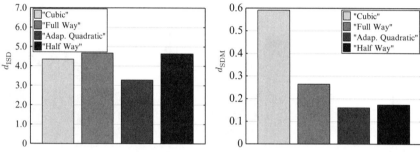

(c) Mean Log spectral distance for low-band excitation signal extension

(d) Mean distance for "Spectral Distortion Measure" for lowband excitation signal extension

Fig. 6.2. Results of the objective quality evaluation for the lower extension region for (**a**) Itakura–Saito, (**b**) Itakura, (**c**) log spectral distortion and (**d**) "Spectral distortion Measure"

speech signals. Each set contained the same utterance processed by the $K = 4$ different algorithms. The subjects could listen to each of the processed files as often as they liked to. Then they had to sort the files by speech quality, starting with the best corresponding to index 1 and ending with the worst corresponding to index 4. This can be described by a voting function

$$V(k, i, m) \in [1...K],\qquad(6.2)$$

with the respective parameters $k \in [1, ..., K]$, $i \in [1, ..., I]$ and $m \in [1, ..., M]$. The mean result of this voting can be calculated as

$$\overline{V}(k) = \frac{1}{IM} \sum_{i=1}^{I} \sum_{m=1}^{M} V(k, i, m),\qquad(6.3)$$

and is depicted in Fig. 6.3. As above in the case of objective quality criteria the adaptive quadratic characteristic performs best for both, the extension of the upper and lower extension region. However, the judgment for the upper

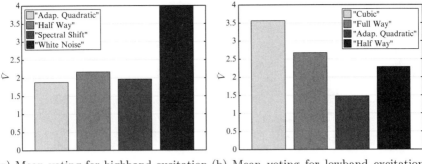

(a) Mean voting for highband excitation signal extension (b) Mean voting for lowband excitation signal extension

Fig. 6.3. Votings for (a) highband and (b) lowband excitation signal extension

extension region is very close for the adaptive quadratic characteristic, the half way rectification, and the spectral shifting approach. Only the approach using a white noise generator performs significantly worse. For the lower extension region the results differ more. A significance analysis of the subjective listening test is presented below.

Figure 6.4 shows the distribution of the votes for the upper extension region over the possible ranks. The distribution of the votes for the lower extension region is shown in Fig. 6.5. These figures are very interesting since for the extension of the upper frequency band for example the spectral shifting method has more votes rating it the best method than the overall winner, the adaptive quadratic characteristic. Whereas the adaptive quadratic characteristic has way more votes rating it second best. The spectral shifting approach produced sometimes noticeable and bothersome artifacts which reflects in the also immense amount of votes rating the method third best. Only the approach using a white noise generator is permanently voted worse.

For the extension of the lower frequency band the result is more clear. Here the adaptive quadratic characteristic outperforms the other approaches obviously. Also the second best method can clearly be assessed as the half-way rectification. The following approaches in descending order concerning quality issues are the full-way rectification and the cubic characteristic. Also here the results are very distinct.

Significance Analysis of the Subjective Results

An interesting question considering the subjective listening tests is if the results are reliable. One conventional method of statistical analysis to prove whether the results are reliable is the analysis of the statistical significance [Hänsler 04]. For simplicity reasons we will reduce the problem to a two-level test. This can be done for example by examining the winning method. Let us define:

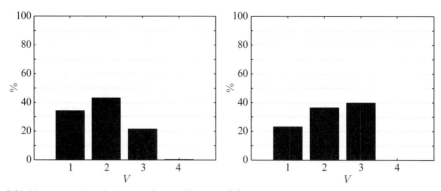

(a) Voting distribution for "Adap. Quadratic"

(b) Voting distribution for "Half Way"

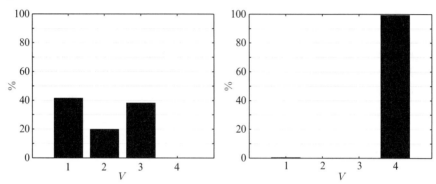

(c) Voting distribution for "Spectral Shift"

(d) Voting distribution for "White Noise"

Fig. 6.4. Distribution of votings for highband for (**a**) "Adap. Quadratic", (**b**) "Half Way", (**c**) "Spectral Shift" and (**d**) "White Noise"

a_-	Number of results voting the winning method second best or worse.
a_+	Number of results voting the winning method best.

This makes a total number of tests

$$N = MI = a_+ + a_-. \tag{6.4}$$

If the tests are mutually independent we can express the probability of actually achieving $\bar{a}_+$ *positive* and $\bar{a}_-$ *negative* results as

$$p\big((a_+ = \bar{a}_+) \wedge (a_- = \bar{a}_-)\big) = \binom{N}{\bar{a}_+} p_+^{\bar{a}_+} p_-^{\bar{a}_-}, \tag{6.5}$$

with the probability p_+ of voting the winning method "best" and p_- of voting the winning method other than "best". Since the sum of both probabilities is

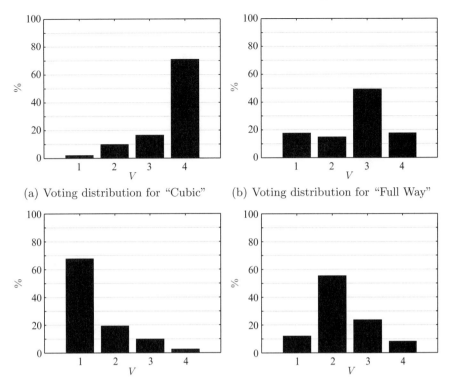

(a) Voting distribution for "Cubic"   (b) Voting distribution for "Full Way"

(c) Voting distribution for "Adap. (d) Voting distribution for "Half Way"
Quadratic"

Fig. 6.5. Distribution of votings for lowband for (**a**) "Cubic", (**b**) "Full Way", (**c**) "Adap. Quadratic" and (**d**) "Half Way"

equal to one we can further write

$$p\left(a_+ = \overline{a}_+\right) = \binom{N}{a_+} p_+^{\overline{a}_+} \left(1 - p_+\right)^{(N - \overline{a}_+)}. \tag{6.6}$$

These probabilities also sum up to one

$$\sum_{k=0}^{N} p(a_+ = k) = \sum_{k=0}^{N} \binom{N}{k} p_+^k \left(1 - p_+\right)^{(N-k)} = 1, \tag{6.7}$$

so that we can write

$$p\left(a_+ = \overline{a}_+\right) = \frac{\binom{N}{a_+} p_+^{\overline{a}_+} \left(1 - p_+\right)^{(N - \overline{a}_+)}}{\sum\limits_{k=0}^{N} \binom{N}{k} p_+^k \left(1 - p_+\right)^{(N-k)}}. \tag{6.8}$$

Now we can compute the probability for achieving at least $\bar{a}_+$ *positive* results

$$p(a_+ \geq \bar{a}_+) = \sum_{k=\bar{a}_+}^{N} p(a_+ = k)$$

$$= \frac{\sum_{k=\bar{a}_+}^{N} \binom{N}{k} p_+^k (1 - p_+)^{(N-k)}}{\sum_{k=0}^{N} \binom{N}{k} p_+^k (1 - p_+)^{(N-k)}}$$

$$= \frac{\sum_{k=\bar{a}_+}^{N} \binom{N}{k} \left(\frac{p_+}{1-p_+}\right)^k}{\sum_{k=0}^{N} \binom{N}{k} \left(\frac{p_+}{1-p_+}\right)^k}. \tag{6.9}$$

By introducing the abbreviation

$$r = \frac{p_+}{p_-} = \frac{p_+}{1 - p_+}, \tag{6.10}$$

and by using the definition of the binomial coefficient

$$\binom{N}{k} = \frac{N!}{k! (N - k)!}, \tag{6.11}$$

we can shorten (6.9) to

$$p(a_+ \geq \bar{a}_+) = \frac{\sum_{k=\bar{a}_+}^{N} \frac{r^k}{k!(N-k)!}}{\sum_{k=0}^{N} \frac{r^k}{k!(N-k)!}}. \tag{6.12}$$

Based on (6.12) we will now derive an upper limit for achieving at least $\bar{a}_+$ *positive* results under the following assumptions:

Hypothesis H_0:   Assuming that $p_+ \leq p_-$, meaning $p_+ \leq \frac{1}{2}$.
Hypothesis H_1:   Assuming that $p_+ > p_-$, meaning $p_+ > \frac{1}{2}$.

Now we would like to compute an upper limit for the conditional probability

$$p\left(a_+ \geq \bar{a}_+\right)\Big|_{H_0} \leq p_0. \tag{6.13}$$

If this upper limit proves to be sufficiently small we can discard hypothesis H_0 and vice versa. For $r \geq 1$ we can specify such an upper bound

$$p\left(a_+ \geq \overline{a}_+\right)\Big|_{p_+ \leq p_-} \leq p\left(a_+ \geq \overline{a}_+\right)\Big|_{p_+ = p_-} = \frac{\sum\limits_{k=\overline{a}_+}^{N} \frac{1}{k!(N-k)!}}{\sum\limits_{k=0}^{N} \frac{1}{k!(N-k)!}} = p_0. \qquad (6.14)$$

Since the computation of the binomial coefficient in (6.14) can lead to numerical inaccuracies, for large N, the binomial distribution is often approximated using a normal distribution of appropriate mean and variance. This approximation leads to

$$p_0\left(\overline{a}_+, N\right)\Big|_{N \gg 1} \approx 1 - \Phi\left(\frac{2\overline{a}_+ - N}{\sqrt{N}}\right), \qquad (6.15)$$

where

$$\Phi(x) = \frac{1}{\sqrt{2\pi}} \int_{-\infty}^{x} e^{-\frac{1}{2}t^2} dt. \qquad (6.16)$$

Now we can finally derive an upper bound for the probability of achieving $\overline{a}_+$ or more *positive* results under the condition that hypothesis H_0 is true. If we examine for example the probability of achieving $\overline{a}_+ = 122$ results, voting the extension of the excitation signal for the lower extension region using the adaptive quadratic characteristic "best", under the assumption that the probability for a person voting the signal extended by the adaptive quadratic characteristic "best" is smaller than the probability that the signal extended by the adaptive quadratic characteristic is voted worse than "best" $p_+ \leq p_-$, we achieve a probability of $p_0(122, 180) \approx 0.0013$. Since this probability is rather small, we can discard the hypothesis that the method of the adaptive quadratic characteristic does not produce the best results considering the extension of the lower extension region. Table 6.1 shows the probability p_0 for the different methods applied to the different extension regions and evaluated for two different kind of groupings. The first grouping counts only the results

Table 6.1. Analysis of the probabilities p_0 for the different methods for extending the upper and lower extension band. The table shows two different groupings of *positive* results. The first contains only the amount of number one votings and the second one contains the amount of number one and number two votings combined in one group and compared with the other votings

Probability p_0 for different methods and result groupings							
		Evaluated algorithm					
		Cubic	Full way	Half way	Adap. quad.	Spectral shift	White noise
a_+, "1"	Low	1.0000	1.0000	1.0000	0.0013		
	High			1.0000	0.9987	0.9033	1.0000
a_+, "1" $\wedge$ "2"	Low	1.0000	1.0000	0.0013	0.0000		
	High			0.0967	0.0000	0.0968	1.0000

labeled with "1" for a_+ and the second grouping counts the results labeled with "1" and with "2" for a_+. The test itself is still a two level test. Only for the second grouping now we add up the results voting the method best and second best for a_+. In both extension regions the method of the adaptive quadratic characteristic shows reliable *positive* results. Note that not all entries are completed due to the evaluation only of a preselection of promising results for each extension region.

For the extension of the lower frequency band the adaptive quadratic characteristic shows significant results, since the probability that the approach is voted best, although it is not the best method, is rather small already for using only the votes for a_+ rating the approach best.

For the upper extension region the evaluation using only votes for a_+ rating the approach best does not indicate any significant result. But when using the votes rating the approach best or second best are summed up for a_+ two methods show significant improvement. Again the adaptive quadratic characteristic performs very well as well as the half-way rectification. However the adaptive quadratic characteristic shows slightly more significant results, since the probability runs towards zero.

Since the objective as well as the subjective evaluation schemes for the extension of the excitation signal show unanimously that the adaptive quadratic characteristic is the best choice for this task it has been used for the implementation of a demonstrator.

6.1.3 Rank Correlation Between Objective and Subjective Results

Since the most important quality criterion is the perceptual quality, the subjective listening tests are most important. Unfortunately these tests are very time consuming, expensive and are not suitable for testing a large series of small parameter changes. Therefore, the need for an instrumental quality criterion which correlates with a subjective listening test arises. For evaluating the correlation between the above presented distance measures and the subjective listening test a rank correlation analysis has been employed. By defining the mean voting for a single signal over all subjects within the listening test as

$$\overline{V}(k,i) = \frac{1}{M} \sum_{m=1}^{M} V(k,i,m), \qquad (6.17)$$

and by introducing a ranking function that returns the position n of a specific element $x(\ell)$ in the sorted list of all L elements

$$\mathcal{R}\big(x(\ell)\big) = n, \text{ for } \overbrace{\underbrace{x(...) < x(...) < ... < x(...)}_{n-1} < x(\ell) < ... < x(...)}^{L}, \qquad (6.18)$$

we can finally compute the rank correlation between the subjective listening test and the distance measure d

Table 6.2. Mean rank correlation between the different objective quality criteria and the subjective quality criterion

Distance measure	Mean rank correlation $\bar{\rho}$	
	Extension of the frequency band	
	0–300 Hz	3.4–5.5 kHz
Itakura (d_I)	0.28	0.48
Itakura–Saito (d_{IS})	−0.09	0.49
LSD (d_{LSD})	0.29	0.51
SDM (d_{SDM})	0.57	0.67

$$\rho(i) = \frac{\sum\limits_{k=1}^{K}\left(\mathcal{R}\big(\overline{V}(k,i)\big)-\mathcal{E}\big\{\mathcal{R}\big(\overline{V}(k,i)\big)\big\}\right)\left(\mathcal{R}\big(d(k,i)\big)-\mathcal{E}\big\{\mathcal{R}\big(d(k,i)\big)\big\}\right)}{\sqrt{\sum\limits_{k=1}^{K}\left(\mathcal{R}\big(\overline{V}(k,i)\big)-\mathcal{E}\big\{\mathcal{R}\big(\overline{V}(k,i)\big)\big\}\right)^2\sum\limits_{k=1}^{K}\left(\mathcal{R}\big(d(k,i)\big)-\mathcal{E}\big\{\mathcal{R}\big(d(k,i)\big)\big\}\right)^2}}. \tag{6.19}$$

Then we can compute the mean rank correlation by a summation over all signals used within the tests

$$\bar{\rho} = \frac{1}{I}\sum_{i=1}^{I}\rho(i). \tag{6.20}$$

The rank correlation reaches values from -1 to 1, where $\bar{\rho} = 0$ means that there is no correlation between the subjective and the objective quality criteria. Tab. 6.2 shows the results of the rank correlation between the different objective quality criteria and the subjective one as above separately evaluated for the upper and the lower extension region. The spectral distortion measure as introduced within this book in Sect. 3.4.7 shows the best results. However the absolute correlation still is not very distinctive meaning that more complex methods that try to simulate the human perception should be applied for future evaluation tasks.

Another aspect of this consideration is that the conventional objective quality criteria are not reliable for this task. This means that listening test are inevitable for quality assessment concerning bandwidth extension or at least the excitation signal extension within an algorithm for bandwidth extension.

6.2 Evaluation of the Envelope Estimation

The evaluation of the envelope estimation has been executed in two stages. In the first stage the most promising parameter set has been determined and in the second stage this method has been improved. Table 6.3 shows a short summary of the different used parameter sets to constitute the labeling of them. For this first stage of evaluation the codebooks have been trained using the TIMIT speech data base instead of the recordings under real conditions

Table 6.3. Overview on different types of codebooks that have been used

Overview and labeling of the different parameter sets					
	Method A	Method B	Method C	Method D	Method E
Type of coefficients used	LPC	CEPS	CEPS + features	MFCC	CEPS + ACM (two stages approach)
Trained up to a codebook size of	4,096	256	512	256	512
Database used	TIMIT	TIMIT	TIMIT	TIMIT	TIMIT
Comment	Trained on LSFs but converted to LPC coefficients for operation mode		Additional features are ACI, ACM, and HLR	Mel scale using 26 filters	Codebook of size x contains x entries labeled as "voiced" and x entries labeled as "unvoiced" (2x in total)
Distance measure used during training	d_{lsf}	d_{ceps}	$d_{\mathrm{ceps,feat}}$	d_{mfcc}	d_{ceps}
Distance measure used during operation	d_{lpc}	d_{ceps}	$d_{\mathrm{ceps,feat}}$	d_{mfcc}	d_{ceps}
Amount of coefficients used for narrowband	12	18	21	13	18
Amount of coefficients used for broadband	20	30	30	28	30

mentioned before. The TIMIT speech data base has been filtered by a band-pass according to [ITU 88b] and afterwards this signal has been coded and decoded using the *enhanced full-rate coder* according to [ETSI 00] which is currently the most common coder within the GSM standard. All other steps of data preparation followed the description depicted in Fig. 5.4 in Sect. 5.1.

Method A produced codebooks up to a size of 4096 entries. The codebooks of method A have been trained using LSF coefficients. For the operation of the codebook the entries have been transformed into LPC coefficients. This has been done due to the fact that the LSF coefficients have a very docile quantization behavior during the training phase and the LPC coefficients produce less computational load during the operation mode. For the computation of the LSF coefficients the LPC coefficients would have to be determined anyway but additionally the zeros of the mirror polynomials would have to be calculated in real-time (see Sect. 3.2.5). This separation of training phase and operation mode involves the usage of two different distance measures for the training and the operation mode.

Method B involves the usage of cepstral coefficients. This approach was processed up to a codebook size of 256 entries. The distance measure used for training and operation was the standard cepstral distance measure. The TIMIT speech data base has been used for the training as well.

Method C also uses cepstral coefficients. In addition three features (ACI, ACM, HLR) have been used for training the codebook. The distance measure used to produce the codebook is introduced in (5.25). The same distance measure has been used within the operation mode. The codebook has been trained up to a size of 512 entries. As before the TIMIT speech data base has been employed to train the codebooks.

Method D involves the usage of MFCCs. For the generation of the MFCCs 26 mel-filters have been employed. Codebooks up to a size of 256 entries have been trained. The distance measure used within the training phase and the operation mode has been introduced in (5.32) within Sect. 5.3.3. Once again the TIMIT speech data base has been used to train the codebooks.

Method E uses the idea of a preclassification depending on the ACM value to determine whether the actual utterance is voiced or unvoiced. The code-books have been trained using cepstral coefficients and the ordinary cepstral distance measure during the training phase and the operation mode. The codebooks have been trained up to a size of 512 entries meaning that the biggest codebook consists of a codebook containing 512 entries labeled as voiced and 512 entries labeled as unvoiced. This fact has to be kept in mind when comparing the performance of the different codebooks of the "same" size.

The neural network approach presented in Sect. 5.4 has not been part of the objective evaluation (see Sect. 6.2.1) since the prior conducted subjective evaluation (see Sect. 6.2.2) showed disadvantages for this specific implementation compared to the codebook approach. Therefore and for the already mentioned reasons that the neural network is difficult to handle and to interfere (see Sect. 5.4.1) this approach has not been examined any further after the subjective evaluation.

6.2.1 Objective Quality Criteria

To evaluate the quality of the different approaches listed in Table 6.3 on page 108 for the estimation of the broadband spectral envelope, the distance between the original broadband spectral envelope, corresponding to the narrowband input signal of the extension algorithm, and the spectral envelope of the extended output signal, on the basis of the narrowband signal, has been measured. This has been done using a evaluation set that has not been part of the training data set. As distance measures the spectral distortion measures already introduced in Sect. 3.4 have been employed. Since the spectral envelopes stored in the codebooks of the different methods are normalized and do not contain any information on power, a power adjustment of the spectral envelopes within the telephone band corresponding to the method presented in Sect. 4.5 has been performed. As distortion measures the Itakura–Saito distance, the log spectral distortion and the *spectral distortion measure* have been employed. Table 6.4 shows the results of the different methods evaluated using the *spectral distortion measure* over the complete frequency range

Table 6.4. Spectral distance measure (SDM) evaluated for different methods and different codebook sizes over the full frequency range

Performances measured with d_{SDM}					
Codebook size	Method A	Method B	Method C	Method D	Method E
2	31.9607	38.4672	39.3236	35.6187	24.9641
4	21.8327	23.6610	26.4531	23.1429	21.2008
8	16.6717	17.1183	24.0615	18.0217	17.0059
16	14.7113	14.6372	21.1985	15.3264	13.5901
32	13.8704	13.2898	17.8085	14.1007	12.7668
64	13.3915	12.4368	14.9111	13.1759	12.1714
128	12.8138	11.8933	13.7505	12.2393	11.5853
256	13.9031	11.4098	12.8863	11.7616	11.2086

Table 6.5. Spectral distance measure (SDM) evaluated for different methods and different codebook sizes over the full frequency range

Performances measured with d_{SDM}		
Codebook size	Method A + LM	Method B + LM	Method D + LM
2	14.1783	15.3642	23.7341
4	11.0506	11.5423	16.7185
8	9.3690	9.2135	13.7184
16	8.9844	8.7095	12.5448
32	8.5876	8.1036	11.4421
64	8.4092	7.6391	10.8431
128	8.2214	7.3836	10.1974
256	9.2898	7.2285	9.7897

for the different codebook sizes. The results for the other distance measures and additionally evaluated for the different extension regions can be found in Appendix F. Table 6.4 shows that method B produces best results. Note that method E first appears to be the winning method, but we have to keep in mind that the codebook labeled with 256 entries contains 512 entries. This explains the slight advantage of this method. But when comparing the performance of codebooks of equal size, method B performs better.

As mentioned in Sect. 5.6 a very interesting and promising approach is the combination of a preclassification using codebooks and the application of linear mapping matrices after this classification. This has been implemented for methods A, B, and D and the results are listed in Table 6.5. As can be seen method B once again outperforms the other approaches. Another very interesting aspect of the results in Table 6.5 is the fact that using a codebook with four entries and subsequent linear mapping produces the same quality as using a codebook consisting of 256 entries. This means that we can either increase quality and maintain the computational complexity or maintain the quality and decrease the computational complexity drastically.

Since method B outperforms the other approaches, this approach has been used for further improvement. The codebook trainings considered in the following have been accomplished using real transmitted and recorded speech data. These further improvements mentioned, consist of the dismissal of narrowband envelopes that do not correlate in an appropriate manner with their broadband counterparts. Therefore a weighting function has been introduced in the narrowband path of the training algorithm making use of the distance measure introduced in (6.1)

$$w = \begin{cases} 1, \text{ for } d_{\text{LSD},\Omega_{\text{low}},\Omega_{\text{high}}} (A_{\text{bb}}, A_{\text{nb}}) < \mu \\ 0, \text{ otherwise.} \end{cases} \qquad (6.21)$$

This distance measure is evaluated from the lower cut-off frequency Ω_{low} of the telephone bandpass up to the upper cut-off frequency Ω_{high}. Where μ denotes the distance threshold. If this threshold is exceeded the corresponding narrowband envelope is not used within the training material. The broadband counterpart is still used since the envelope is guaranteed to be undistorted. One can think of introducing a weighting function that is not a hard decision as implemented in (6.21) but a weighting function that depends on the distance. The need for such a mechanism is due to the fact that the data before and after the transmission over a GSM system do not correlate anymore sometimes because of bad radio reception or complete drop outs.

Additionally the autocorrelation modification presented in Sect. 5.2 has been included. Table 6.6 shows the results obtained by the modifications evaluated using the spectral distortion measure d_{SDM}. Where the different methods are:

Table 6.6. Spectral distance measure (SDM) evaluated for different methods and different codebook sizes over the full frequency range.

Codebook size	Method I	Method II	Method III	Method IV
	Performances measured with d_{SDM}			
2	23.6306	21.0931	20.7069	22.7029
4	21.4951	18.1437	18.5355	20.0975
8	15.4715	13.1396	12.9973	15.2616
16	12.3869	11.0496	11.0726	12.2711
32	11.6591	9.8693	10.1841	11.2569
64	10.6514	8.9418	9.3197	10.6116
128	10.1179	8.5427	8.7971	9.9724
256	9.6027	7.8145	8.3313	9.1111

Method I	Codebook using the autocorrelation modification but without the distance dependent weighting function (6.21)
Method II	Codebook without the autocorrelation manipulation but with the distance dependent weighting function
Method III	Codebook without autocorrelation manipulation and without distance dependent weighting function.
Method IV	Codebook using the autocorrelation modification and the distance dependent weighting function.

The result might seem unsatisfactory since method IV produces not the best results. But this is due to the fact that for the generation of the training and evaluation data only one and the same telephone bandpass has been used. The application of the autocorrelation manipulation introduces in this case distortion that is not needed. The advantage of this method becomes obvious when applied to a database consisting of recordings using several telephone bandpasses. The fact that method II is better than methods III and IV is better than method I shows that the distance dependent weighting function is important and has a positive influence. The introduction of the distance dependent weighting function prevents distortions introduced by possible drop outs or GSM buzz or a bad radio reception in general. Another aspect is the slight performance drop compared to the distance produced by method B in Table 6.5. This is due to the usage of real transmitted data and therewith the channel variability introduced by the usage of a real GSM system. Results using other distance measures and different evaluation bandwidths can be found in Appendix F.

6.2.2 Subjective Quality Criteria

According to the evaluation of the excitation signal also subjective listening tests have been carried out for the evaluation of the spectral envelope extension. In order to evaluate the subjective quality of the extended spectral envelope a mean-opinion-score (MOS) test in terms of a comparison rating has been executed. About 30 people of different age and gender have participated in the test. The subjects were asked to compare the quality of two signals (pairs of bandlimited and extended signals) by choosing one of the statements listed in Table 6.7. For the generation of the extended excitation signal a cubic characteristic has been used in both extension methods compared in the test. This guarantees that concerning the comparison of the two extension schemes, only the quality of the spectral envelope extension is judged. The methods used for the comparison comprise method A of Table 6.3 on page 108, using the codebook with 1,024 entries, and a neural network approach using cepstral coefficients. The neural network was a multilayer perceptron in feed forward operation as described in Sect. 5.4 consisting of three layers. The input layer

Table 6.7. Conditions of the MOS test for subjective evaluation of the spectral envelope extension

Score	Statement
−3	A is much worse than B
−2	A is worse than B
−1	A is slightly worse than B
0	A and B are about the same
1	A is slightly better than B
2	A is better than B
3	A is much better than B

and the hidden layer contained 36 neurons and the output layer contained 30 neurons. As input two times 18 cepstral coefficients that have been calculated on the basis of 12 LPC coefficients (see (3.84)) have been used. Two times 18 means that the cepstral coefficients of the current and the previous frame have been used to implement a short memory. The narrowband and broadband counterparts of the extended signals have not been part of the training data base used for the neural network or the codebook. Finally, the subjects were asked in a second step whether they prefer the signal which was extended by the neural network or the one which was extended with the codebook. The result of this listening test is presented in Fig. 6.6. The results show clearly that the codebook approach outperforms the neural network approach. Admittedly neural networks of different complexity concerning the amount of used neurons have not been investigated in detail [Zaykovskiy 05]. However regarding the disadvantages of neural networks that have already been mentioned in Sect. 5.4 this delivers another reason for not longer pursuing this approach.

Another aspect of this result is that the votes for both approaches have a local maximum within each half of the voting scale. This means that the subjects either preferred the extended signal or the bandlimited reference signal in a comparison. But they did not vote them equal, which means that even if the codebook approach achieved fair results in average a non negligible amount of subjects disliked the signals produced by this method.

All in all the subjects sometimes had difficulties in finding a difference between the processed and the non-processed signals and in developing a preference for a certain method. This lead to the belief that subjective listening tests for evaluating the quality of spectral envelope extension schemes are not meaningful enough to justify the immense effort associated with the realization of such a test. Therefore further investigations using the other methods introduced in Tab. 6.3 on page 108 have been discarded and only informal listening tests have been performed.

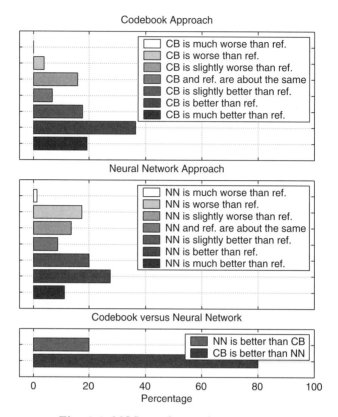

Fig. 6.6. MOS-test for envelope extension

6.3 Discussion

In this chapter we have presented several methods to evaluate the quality of the different parts within a system for extending the bandwidth of speech signals. Different implementations of these parts consisting of the extension of the excitation signal and the extension of the broadband spectral envelope have been evaluated using the presented methods.

For the extension of the excitation signal all applied evaluation measures and tests showed the superiority of the adaptive quadratic characteristic introduced in Sect. 4.2. This justifies the usage of this method within a demonstrator or system for extending the bandwidth of speech signals.

For the extension of the spectral envelope the results are not as clear as for the excitation signal but a trend can be observed too. Method B of Table 6.3 on page 108 shows the best results and is therefore a very interesting candidate in combination with a succeeding linear mapping. The effect of the modification of the autocorrelation coefficients (method IV on page 112) however has to be examined in more detail using data that has been recorded using different realistic transmission scenarios and therewith different bandpass characteristics.

7

Summary

In this book various methods for extending the bandwidth of telephone bandlimited speech signals have been presented, discussed and evaluated. Focus of the book was laid on the problems that arise from dealing with real-world signals. The required basic methods for speech analysis and synthesis have been recapitulated as well as the extraction of different speech features and adequate distance measures including an improved distance measure for a terminal evaluation have been presented. The approaches for extending the bandwidth of a speech signal have been split into the generation of an extended excitation signal and the extension of a spectral envelope following the source-filter model for the speech production process. These parts have been discussed and evaluated separately.

For the generation of the extended excitation signal several methods have been introduced and evaluated in an objective, as well as in a subjective manner. The method that produced best results is the approach using an adaptive quadratic characteristic first introduced in this book. Fortunately the required computational load for this approach is small in comparison to pitch adaptive methods for example. A significance analysis has been performed for the subjective quality evaluation. One of the distance measures used for the objective evaluation has also been derived within the work for this book [Iser 06]. An analysis of the different objective quality measures using the rank correlation between objective and subjective results showed that this distance measure is superior to the established distance measures. However the absolute value of the rank correlation is still rather small and even more complex distance measures simulating the human auditory system should be tested. It should be noted in this context that the well known PESQ method for evaluating speech quality has proven in inchoate tests not to be a well suited measure for bandwidth extended signals. A very important part of the excitation signal extension concerning the resulting speech quality lays in adjusting the appropriate power of the excitation signal. It is also crucial not to have discontinuities in power from one frame to another. The winning method showed convincing properties concerning this issue as well. As for the lower extension region it is essential that the fundamental frequency and the first harmonics are placed in an accurate manner the adaptive quadratic characteristic once again performs very well as it is the case with most of the non-linear characteristics. Another problem arising from the block processing scheme that was used in this work is the necessity of generating a signal that has a continuous phase over different frames. This problem particularly arises when using modulation or spectral shifting approaches. The proposed method which is applied in the time domain behaves uncritically concerning the phase continuity.

For the generation of the extended spectral envelope a selection of different approaches including approaches derived within the work for this book based on codebooks, neural networks [Iser 03] and linear mapping [Zaykovskiy 05] have been analyzed as well . Another important part that has been presented in more detail that is neglected in most of the prior art is the proper preparation of a training data set consisting of recordings of real transmitted data using a GSM network. In parallel to the evaluation of the excitation signal extension different objective distance measures have been used to evaluate the different approaches. The method that showed the best results is the one using cepstral coefficients implicating the usage of the corresponding distance measures for the training phase as well as for the operation mode. During the training the correlation between the narrowband and the broadband envelopes has been observed to be able to compensate for bad radio reception or even drop outs in a real GSM transmission by discarding frames with insufficient correlation between narrowband and broadband features. Additionally the manipulation of the narrowband autocorrelation coefficients, first introduced in this book, has been used for increased robustness against poor sample rate converters or differences in the band limiting means during training and operation of the algorithm. The usage of linear mapping matrices as a post processing stage has proven to be advantageous and also gives the possibility of saving computational complexity while maintaining the resulting quality at the same time. The algorithm also provides the possibility of using the spectral envelopes of the broadband codebook if the linear mapping fails or passes back unstable filter coefficients. In addition the observance of the distance of the chosen entry of the narrowband codebook was used to turn off the extension if the algorithm runs the risk of degrading the speech quality. The algorithm for the extension of the spectral envelope turned out to be the quality limiting element in the overall system. Additionally this is the part that occupies the lion's share of the computational load. To be more precise the evaluation of the distance measure during the search for the codebook entry producing minimum distance is responsible for the computational load. Another crucial part considering the resulting speech quality is the power adjustment of the spectral envelope coming from the codebook or after the post processing by a linear mapping matrix. This part of the algorithm is responsible for most of the audible artifacts in the extended speech signal. This has been proven by informal listening tests using an extended spectral envelope coming from one of the presented methods but determining the gain for the envelope using the original broadband signal. This setup produced pleasant sounding results. In general it should be noted that all algorithms for extending the spectral envelope have difficulties placing a strong formant in the upper extension region. The reason for this is the absence of a non-ambiguous indicator in the telephone band for a strong formant outside the telephone band in some cases. This means that lots of broadband spectral envelopes in the training data set exist that do look the same within the telephone band but are different in the upper extension region. If these spectral envelopes are put together into a cluster the strong formant in the upper extension region will be downsized through averaging effects.

The algorithm derived within this book as well as a elementary version which has reduced computational complexity have been implemented in a real-time framework and can be demonstrated. The algorithm with reduced complexity has been integrated on a target platform which is used in a car as well. This algorithm is ready for mass-produce and is also planned to be introduced in the market in the very near future. The achievable gain in speech quality for the algorithm with reduced quality is smaller than the gain achievable with the original algorithm. However the elementary approach produces fewer artifacts. The computational load produced by the original approach is despite the reduction obtained by the use of a small codebook with linear mapping matrices as a post processing stage still too much to be integrated commercially.

Another aspect of the work is the question for the use of the derivation of an algorithm extending the bandwidth from approximately 3.1 kHz to approximately 5.5 kHz under the awareness that sooner or later broadband telephony with a bandwidth of approximately 7 kHz will be available by the introduction of the *AMR wideband* codec in mobile communication or the respective codecs for VoIP (voice over IP) or the codec introduced in [ITU 88a] that possesses 7 kHz bandwidth for the fixed land line networks. Experiments using a *narrowband* signal with a bandwidth of 11 kHz and extending this signal with a version of the algorithm described within this book that has been trained for this task to a bandwidth of 22 kHz showed that the speech quality improvement is still remarkable but the artifacts produced by the algorithm are almost gone since the classification of the spectral envelope is considerably easier since more bandwidth and therefore more information on the spectral envelope is available. The above described problem of strong formants being downsized by averaging effects is reduced due to the fact that more bandwidth is available to distinguish the different spectral envelopes. This aspect makes the algorithm derived within this book even more attractive for the future.

For current narrowband systems an interesting alternative for increasing the achievable speech quality of bandwidth extension algorithms is the transmission of hidden information within the audio stream. This technique is called watermarking. First approaches can be found in [Geiser 05]. Using this method it is possible to transmit information that is speaker specific or the terminal on the far side which is able to track the broadband speech can compute correction factors for a common codebook to increase the performance. Disadvantage of these approaches is the increased computational load and the needed initial training phase during the first seconds of a call in the case of computing codebook correction factors.

An additional field of application for the algorithms presented in this book are model-based noise reduction or speech reconstruction schemes. Classical approaches like Wiener filtering lack the possibility of reducing non-stationary noise. If the speech signal itself can be estimated sufficiently accurate, using the presented methods, even non-stationary noise or strong frequency selective perturbances can be reduced. The non-perturbed or only moderately noise afflicted speech parts can be used for estimating the clean speech signal. First approaches for this new field of application have been implemented and showed depending on the kind and intensity of perturbance promising results. Further investigations can be found in [Krini 07] for example.

Within the work for this book publications emerged investigating codebook and neural network approaches for the extension of the spectral envelope [Iser 03] as well as a comparison of neural networks and linear mapping [Zaykovskiy 05]. The methods that have been investigated for the extension of the excitation signal and the results have been published in [Iser 05b]. Descriptions and results of different analyzed overall systems as well as some minor details can be found in [Iser 05a, Iser 06, Iser 08].

A

Levinson–Durbin Recursion

As shown in Sect. 3.1.1 the solution for finding the optimal prediction coefficients $\mathbf{a} = [a_1, a_2, ..., a_P]^{\mathrm{T}}$ of a predictor of order P is given by

$$\mathbf{a} = \mathbf{R}_{ss}^{-1} \mathbf{r}_{ss}, \tag{A.1}$$

with the vector

$$\mathbf{r}_{ss} = [r_{ss}(1), r_{ss}(2), ..., r_{ss}(P)]^{\mathrm{T}} \tag{A.2}$$

and the matrix given by

$$\mathbf{R}_{ss} = \begin{bmatrix} r_{ss}(0) & r_{ss}(1) & \cdots r_{ss}(P-1) \\ r_{ss}(1) & r_{ss}(0) & \cdots r_{ss}(P-2) \\ \vdots & \vdots & \vdots \\ r_{ss}(P-1) & r_{ss}(P-2) & \cdots & r_{ss}(0) \end{bmatrix} \tag{A.3}$$

involving elements given by the estimated short-term autocorrelation function

$$r_{ss}(m) = \sum_{k=0}^{N-1-m} s(k)s(k+m) \tag{A.4}$$

of a speech segment $[s(0), s(1), ..., s(N-1)]$ of length N. Computing the solution of (A.1) by inverting the matrix $\mathbf{R}_{ss}$ without exploiting the special properties of this matrix results in P^3 operations. In addition the limited precision of the elements of $\mathbf{R}_{ss}$ and the limited accuracy of the processing unit may cause numerical problems when inverting the matrix. Therefore methods have been developed that take advantage of the special properties of this matrix. The method presented here is called *Levinson–Durbin recursion*. The Levinson–Durbin recursion is a recursive-in-model-order solution for solving (A.1). This means that the coefficients of an order P predictor are calculated out of the solution for an order $P-1$ predictor. By exploiting the Toeplitz structure of $\mathbf{R}_{ss}$ the Levinson–Durbin recursion reduces the computational complexity to P^2 operations [Hänsler 01].

Let us recall (A.1) and denote the order by adding a superscript in brackets

$$\mathbf{R}_{ss}^{(P)}\mathbf{a}^{(P)} = \mathbf{r}_{ss}^{(P)}. \tag{A.5}$$

We can denote the system of equations defined in (A.5) as

$$
\begin{array}{ccccccc}
r_0 a_1^{(P)} & + & r_1 a_2^{(P)} & + \cdots + & r_{P-1} a_P^{(P)} & = & r_1 \\
\vdots & & \vdots & & \vdots & & \vdots \\
r_{P-2} a_1^{(P)} & + r_{P-3} a_2^{(P)} & + \cdots + & r_1 a_P^{(P)} & & = & r_{P-1} \\
r_{P-1} a_1^{(P)} & + r_{P-2} a_2^{(P)} & + \cdots + & r_0 a_P^{(P)} & & = & r_P.
\end{array}
\tag{A.6}
$$

By subtracting the last expression this system of equations can now be split into a system of equations containing the first $P-1$ rows

$$
\begin{array}{ccccccc}
r_0 a_1^{(P)} & + & r_1 a_2^{(P)} & + \cdots + & r_{P-2} a_{P-1}^{(P)} & = & r_1 - r_{P-1} a_P^{(P)} \\
\vdots & & \vdots & & \vdots & & \vdots \\
r_{P-2} a_1^{(P)} & + r_{P-3} a_2^{(P)} & + \cdots + & r_0 a_{P-1}^{(P)} & & = & r_{P-1} - r_1 a_P^{(P)},
\end{array}
\tag{A.7}
$$

and an equation containing the last row

$$r_{P-1} a_1^{(P)} + r_{P-2} a_2^{(P)} + \cdots + r_1 a_{P-1}^{(P)} = r_P - r_0 a_P^{(P)}. \tag{A.8}$$

Now the system of equations in A.7 contains on the left side the matrix $\mathbf{R}^{(P-1)}$. This matrix is obtained by skipping the last row and the last column of $\mathbf{R}^{(P)}$. By defining

$$\tilde{\mathbf{r}}^{(P)} = [r_{ss}(P), r_{ss}(P-1), ..., r_{ss}(1)]^{\mathrm{T}}, \tag{A.9}$$

the system of equations A.7 can be written as

$$\mathbf{R}^{(P-1)}
\begin{bmatrix}
a_1^{(P)} \\
a_2^{(P)} \\
\vdots \\
a_{P-1}^{(P)}
\end{bmatrix}
= \mathbf{r}^{(P-1)} - a_P^{(P)} \tilde{\mathbf{r}}^{(P-1)}.
\tag{A.10}$$

In the same manner we can rewrite (A.8):

$$\left(\tilde{\mathbf{r}}^{(P-1)} \right)^{\mathrm{T}}
\begin{bmatrix}
a_1^{(P)} \\
a_2^{(P)} \\
\vdots \\
a_{P-1}^{(P)}
\end{bmatrix}
= r_P - a_P^{(P)} r_0.
\tag{A.11}$$

Now we multiply (A.10) with $\mathbf{R}^{(P-1)^{-1}}$, which is the inverse of $\mathbf{R}^{(P-1)}$:

$$\begin{bmatrix} a_1^{(P)} \\ a_2^{(P)} \\ \vdots \\ a_{P-1}^{(P)} \end{bmatrix} = \mathbf{R}^{(P-1)^{-1}} \mathbf{r}^{(P-1)} - a_P^{(P)} \mathbf{R}^{(P-1)^{-1}} \tilde{\mathbf{r}}^{(P-1)}. \tag{A.12}$$

If we now define in parallel to $\tilde{\mathbf{r}}^{(P)}$

$$\tilde{\mathbf{a}}^{(P)} = \left[a_P^{(P)}, a_{P-1}^{(P)}, ..., a_1^{(P)} \right]^{\mathrm{T}}, \tag{A.13}$$

exploiting the symmetry of A.1

$$\mathbf{R}^{(P)} \tilde{\mathbf{a}}^{(P)} = \tilde{\mathbf{r}}^{(P)}, \tag{A.14}$$

we can write (A.12) as:

$$\begin{bmatrix} a_1^{(P)} \\ a_2^{(P)} \\ \vdots \\ a_{P-1}^{(P)} \end{bmatrix} = \mathbf{a}^{(P-1)} - a_P^{(P)} \tilde{\mathbf{a}}^{(P-1)}. \tag{A.15}$$

This finally is the instruction how to calculate the coefficients $a_i^{(P)}$, $i = 1, ..., P - 1$, of a predictor of order P out of the coefficients of an order $P - 1$ predictor and $a_P^{(P)}$. For calculating the single coefficients (A.15) can be written as:

$$\boxed{a_i^{(P)} = a_i^{(P-1)} - a_P^{(P)} a_{P-i}^{(P-1)} \text{ for } i = 1, ..., P - 1.} \tag{A.16}$$

This is known as the Levinson-recursion. The coefficient that is missing up to now $a_P^{(P)}$ is called reflection coefficient (see (3.55) in Sect. 3.2.1) or PARCOR coefficient (partial correlation coefficient). Using (A.15) in (A.11) results in

$$a_P^{(P)} \left(r_0 - \left(\tilde{\mathbf{r}}^{(P-1)} \right)^{\mathrm{T}} \tilde{\mathbf{a}}^{(P-1)} \right) = r_P - \left(\tilde{\mathbf{r}}^{(P-1)} \right)^{\mathrm{T}} \mathbf{a}^{(P-1)}. \tag{A.17}$$

Now we can isolate $a_P^{(P)}$:

$$\boxed{a_P^{(P)} = \frac{r_P - \left(\tilde{\mathbf{r}}^{(P-1)} \right)^{\mathrm{T}} \mathbf{a}^{(P-1)}}{r_0 - \left(\tilde{\mathbf{r}}^{(P-1)} \right)^{\mathrm{T}} \tilde{\mathbf{a}}^{(P-1)}}.} \tag{A.18}$$

Using (A.16) and (A.18) all coefficients of an order P predictor can be calculated out of an order $P - 1$ predictor with the autocorrelation values of the input process or the estimated autocorrelation values.

Another useful parameter that can also be computed in a recursive manner is the short-term energy of the prediction error E_{error}. This can be helpful if no target order is given. Then the predictor error energy can be observed and if the desired energy is reached the algorithm can be aborted. We can formulate for the short-term energy of the predictor error according to (3.21) using the present notation

$$E_{error}^{(P)} = r_0 - \mathbf{r}^{(P)\,T}\,\mathbf{a}^{(P)}\bigg|_{opt}. \tag{A.19}$$

We can once again isolate $a_P^{(P)}$

$$E_{error}^{(P)} = r_0 - r_P\,a_P^{(P)} - \mathbf{r}^{(P-1)\,T} \begin{bmatrix} a_1^{(P)} \\ a_2^{(P)} \\ \vdots \\ a_{P-1}^{(P)} \end{bmatrix}. \tag{A.20}$$

If we now use (A.15) we get

$$E_{error}^{(P)} = r_0 - \mathbf{r}^{(P-1)\,T}\,\mathbf{a}^{(P-1)} - a_P^{(P)}\left(r_P - \mathbf{r}^{(P-1)\,T}\,\tilde{\mathbf{a}}^{(P-1)}\right). \tag{A.21}$$

Using (A.18) and (A.19) as well as the fact that $\tilde{\mathbf{r}}^{(P-1)\,T}\,\tilde{\mathbf{a}}^{(P-1)} = \mathbf{r}^{(P-1)\,T}\,\mathbf{a}^{(P-1)}$ and $\tilde{\mathbf{r}}^{(P-1)\,T}\,\mathbf{a}^{(P-1)} = \mathbf{r}^{(P-1)\,T}\,\tilde{\mathbf{a}}^{(P-1)}$ we finally get a recursive rule for computing the energy of the predictor error signal

$$\boxed{E_{error}^{(P)} = E_{error}^{(P-1)}\left(1 - a_P^{(P)\,2}\right).} \tag{A.22}$$

The energy of the predictor error signal is guaranteed to decrease monotonously over the order of the predictor filter. If this is not the case in an implementation this is due to numerical problems or inaccuracies.

One important thing that is still missing up to now are the initial values for starting the recursion that emerge from the trivial solution of no prediction. They can be characterized by

$$\boxed{\begin{aligned} a_0^{(0)} &= 1, \\ E_{error}^{(0)} &= r_0. \end{aligned}} \tag{A.23}$$

B

LBG-Algorithm

In this section we would like to introduce the basic LBG-algorithm as first published by [Linde 80] and named after its inventors Linde, Buzo and Gray. The LBG-algorithm is an efficient and intuitive algorithm for vector quantizer design based on a long training sequence of data. Various modifications exist (see [LeBlanc 93, Paliwal 93] for example). In this book the LBG-algorithm is used for the generation of a codebook containing the spectral envelopes that are most representative in the sense of a distance measure for a given set of training data. For the generation of this codebook the following iterative procedure is applied to the training data:

1. Initializing:
   Compute the centroid for the whole training data. The centroid is defined as the vector with minimum distance in the sense of a distortion measure to the complete training data.

2. Splitting:
   Each centroid is splitted into two near vectors by the application of a perturbance.

3. Quantization:
   The whole training data is assigned to the centroids by the application of a certain distance measure and afterwards the centroids are calculated again. Step 3 is executed again and again until the result does not show any significant changes. Is the desired codebook size reached ⇒ abort. Otherwise continue with step 2.

Figure B.1 shows the functional principle of the LBG-Algorithm by applying the algorithm to a training data consisting of two clustering points (see Fig. B.1a). Only one iteration is depicted. Starting with step 1 the centroid over the whole training data is calculated which is depicted in Fig. B.1b. In step 2 this centroid is split into two near initial centroids by the application

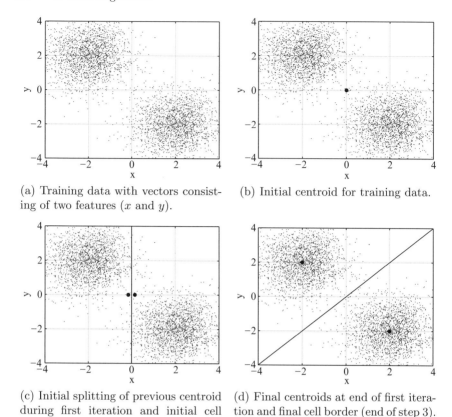

(a) Training data with vectors consisting of two features (x and y).

(b) Initial centroid for training data.

(c) Initial splitting of previous centroid during first iteration and initial cell border (step 2).

(d) Final centroids at end of first iteration and final cell border (end of step 3).

Fig. B.1. Example for the vector quantization of a training data with two clustering points following the LBG-algorithm

of a perturbance as can be seen in Fig. B.1c. Afterwards in step 3 the training data is quantized to the new centroids and after quantization the new centroids are calculated. This procedure is repeated until a predefined overall distortion is reached or a maximum amount of iterations or no significant changes occur. This final state of the first iteration is depicted in Fig. B.1d. Using the LBG-algorithm this way, it is only possible to obtain codebooks with an amount of entries equal to a power of two. This could be circumvented by either splitting in more than just two new initial centroids or by starting with an initial guess of the desired n centroids or codebook entries, respectively and quantizing them as long as the centroids change considerably. Another possible method consists of splitting only a subset of the existing centroids. This could be done by observing the mean distance of the features classified to one centroid and splitting only those whose features produce the worst mean distance.

C

Generalized Delta Rule or Standard Backpropagation

In the following we will give a short introduction of the training algorithm used within this book to train the neural network or to be more specific the multilayer perceptron in feed forward operation [Rojas 96]. Learning the pattern or functionality of a forced learning task $\tilde{\mathcal{L}}$, consisting of

$$p \in \{0, ..., N_{\text{pattern}} - 1\} \tag{C.1}$$

pairs of input patterns

$$\mathbf{i}^{(p)} = \left[i_0^{(p)}, i_1^{(p)}, ..., i_{N_{\nu=0}-1}^{(p)} \right]^{\mathrm{T}} \tag{C.2}$$

and target patterns

$$\mathbf{t}^{(p)} = \left[t_0^{(p)}, t_1^{(p)}, ..., t_{N_{\nu=N_{\text{layer}}-1}-1}^{(p)} \right]^{\mathrm{T}}, \tag{C.3}$$

starts with the computation of an appropriate distance measure, measuring the distance between the actual output of the neural network and the desired target value. The error signal E obtained by the application of an appropriate distance measure is propagated backwards through the net starting from the output layer towards the input layer. This assures the possibility for the inner neurons to determine their own error part. The error of all units then finally is responsible for the modification of the individual weights. This method is known as backpropagation algorithm or generalized delta rule and can be annotated as

$$\Delta_p \, w_{\mu,\nu-1,\lambda} = \eta \, \delta_{\lambda,\nu}^{(p)} \, a_{\mu,\nu-1}^{(p)}, \tag{C.4}$$

with

$$\mu \in \{0, ..., N_{\nu-1} - 1\} \tag{C.5}$$

being the index representing the node in layer $\nu - 1$,

$$\nu \in \{1, ..., N_{\text{layer}} - 1\} \tag{C.6}$$

representing the actual layer,

$$\lambda \in \{0, ..., N_\nu - 1\} \tag{C.7}$$

being the index representing the node in the actual layer ν, $\eta > 0$ being the so-called *learning rate* and finally p being one learning pattern of the forced learning task $\tilde{\mathcal{L}}$. Furthermore $\delta_{\lambda,\nu}^{(p)}$ is defined as

$$\delta_{\lambda,\nu}^{(p)} = \begin{cases} f'_{\text{act}}(u_{\lambda,\nu}^{(p)})(t_\lambda^{(p)} - a_{\lambda,\nu}^{(p)}), & \text{for } \lambda \in \{0, ..., N_\nu - 1\}, \\ & \nu = N_{\text{layer}} - 1, \\ f'_{\text{act}}(u_{\lambda,\nu}^{(p)}) \sum_{\kappa=0}^{N_{\nu+1}-1} \delta_{\kappa,\nu+1}^{(p)} \, w_{\lambda,\nu,\kappa}, & \text{for } \lambda \in \{0, ..., N_\nu - 1\}, \\ & \nu \in \{1, ..., N_{\text{layer}} - 2\}, \end{cases} \tag{C.8}$$

with $u_{\lambda,\nu}^{(p)}$ being the input of node λ in layer ν

$$u_{\lambda,\nu}^{(p)} = \theta_{\lambda,\nu} + \sum_{\mu=0}^{N_{\nu-1}-1} a_{\mu,\nu-1}^{(p)} \, w_{\mu,\nu-1,\lambda} \tag{C.9}$$

and $w_{i,\nu-1,\lambda}$ representing the associated weights as well as $\theta_{\lambda,\nu}$ the associated bias values. The output of a node is given by

$$a_{\lambda,\nu}^{(p)} = f_{\text{act}}\left(u_{\lambda,\nu}^{(p)}\right). \tag{C.10}$$

In the following we will briefly describe the derivation of this rule for the modification of the weights. As a cost function we will define the sum of the squared differences of the desired target and the actual output of the neural network

$$E = \sum_{p=0}^{N_{\text{pattern}}-1} E^{(p)} = \frac{1}{2} \sum_{p=0}^{N_{\text{pattern}}-1} \sum_{\lambda=0}^{N_\nu-1} (t_\lambda^{(p)} - a_{\lambda,\nu}^{(p)})^2, \tag{C.11}$$

for $\nu = N_{\text{layer}} - 1$. Aim of the algorithm is to minimize the error E by modifying the weights $w_{\mu,\nu-1,\lambda}$. Looking for a minimum by variation of the weights means

$$\frac{\partial E^{(p)}}{\partial w_{\mu,\nu-1,\lambda}} = 0. \tag{C.12}$$

When looking at (C.11) we can state that the error is a function of the output $a_{\lambda,\nu}^{(p)}$. The output in turn is a function of the input and the weights $w_{\mu,\nu-1,\lambda}$. So we can further write using the chain rule

$$\frac{\partial E^{(p)}}{\partial w_{\mu,\nu-1,\lambda}} = \frac{\partial E^{(p)}}{\partial u_{\lambda,\nu}^{(p)}} \frac{\partial u_{\lambda,\nu}^{(p)}}{\partial w_{\mu,\nu-1,\lambda}}. \tag{C.13}$$

With (C.9) we get for the second factor in (C.13)

$$\frac{\partial u_{\lambda,\nu}^{(p)}}{\partial w_{\mu,\nu-1,\lambda}} = \frac{\partial}{\partial w_{\mu,\nu-1,\lambda}} \sum_{\mu=0}^{N_{\nu-1}-1} a_{\mu,\nu-1}^{(p)} w_{\mu,\nu-1,\lambda} + \theta_{\lambda,\nu} = a_{\mu,\nu-1}^{(p)}. \quad (C.14)$$

Using the error signal $\delta_{\lambda,\nu}^{(p)}$ (compare (C.8))

$$\delta_{\lambda,\nu}^{(p)} = -\frac{\partial E^{(p)}}{\partial u_{\lambda,\nu}^{(p)}}, \quad (C.15)$$

we can write

$$-\frac{\partial E^{(p)}}{\partial w_{\mu,\nu-1,\lambda}} = \delta_{\lambda,\nu}^{(p)} a_{\mu,\nu-1}^{(p)}. \quad (C.16)$$

For realizing a gradient descend within E the modification of the weights has to follow (C.4). Now $\delta_{\lambda,\nu}^{(p)}$ has to be computed for each node of the network. In the following we will show that a simple recursive algorithm exists where the $\delta_{\lambda,\nu}^{(p)}$ can be computed by propagating the error backwards through the network.

By once again applying the chain rule to (C.15) we obtain the following product:

$$\delta_{\lambda,\nu}^{(p)} = -\frac{\partial E^{(p)}}{\partial u_{\lambda,\nu}^{(p)}} = -\frac{\partial E^{(p)}}{\partial a_{\lambda,\nu}^{(p)}} \frac{\partial a_{\lambda,\nu}^{(p)}}{\partial u_{\lambda,\nu}^{(p)}}. \quad (C.17)$$

For the second factor in (C.17) we can write

$$\frac{\partial a_{\lambda,\nu}^{(p)}}{\partial u_{\lambda,\nu}^{(p)}} = f'_{\text{act}}(u_{\lambda,\nu}^{(p)}). \quad (C.18)$$

Now we can compute the first factor in (C.17) for two different cases:

1. In the first case we consider the output layer $\nu = N_{\text{layer}} - 1$. With (C.11) we can write

$$\frac{\partial E^{(p)}}{\partial a_{\lambda,\nu}^{(p)}} = -(t_\lambda^{(p)} - a_{\lambda,\nu}^{(p)}). \quad (C.19)$$

   Substituting both factors in (C.17) by the expressions of (C.18) and (C.19) we can write for the output layer $\nu = N_{\text{layer}} - 1$

$$\delta_{\lambda,\nu}^{(p)} = f'_{\text{act}}(u_{\lambda,\nu}^{(p)})(t_\lambda^{(p)} - a_{\lambda,\nu}^{(p)}). \quad (C.20)$$

2. In the second case we examine the first factor in (C.17) for $\nu \in \{1, ..., N_{\text{layer}} - 2\}$. By once again applying the chain rule we obtain a sum over the nodes $\kappa \in \{0, ..., N_{\nu+1} - 1\}$ in layer $\nu + 1$ which are connected with node λ in layer ν

$$\frac{\partial E^{(p)}}{\partial a_{\lambda,\nu}^{(p)}} = \sum_{\kappa=0}^{N_{\nu+1}-1} \left(\frac{\partial E^{(p)}}{\partial u_{\kappa,\nu+1}^{(p)}} \frac{\partial u_{\kappa,\nu+1}^{(p)}}{\partial a_{\lambda,\nu}^{(p)}} \right) \tag{C.21}$$

$$= \sum_{\kappa=0}^{N_{\nu+1}-1} \left(\frac{\partial E^{(p)}}{\partial u_{\kappa,\nu+1}^{(p)}} \frac{\partial}{\partial a_{\lambda,\nu}^{(p)}} \sum_{\lambda=0}^{N_{\nu}-1} w_{\lambda,\nu,\kappa} \, a_{\lambda,\nu}^{(p)} + \theta_{\kappa,\nu+1} \right)$$

$$= \sum_{\kappa=0}^{N_{\nu+1}-1} \frac{\partial E^{(p)}}{\partial u_{\kappa,\nu+1}^{(p)}} \, w_{\lambda,\nu,\kappa}$$

$$= - \sum_{\kappa=0}^{N_{\nu+1}-1} \delta_{\kappa,\nu+1}^{(p)} \, w_{\lambda,\nu,\kappa}. \tag{C.22}$$

By replacing the two factors in (C.17) with the expressions obtained in (C.18) and (C.22) we now can express $\delta_{\lambda,\nu}^{(p)}$ for all nodes within the layers $\nu \in \{0, ..., N_{\text{layer}} - 2\}$ as

$$\delta_{\lambda,\nu}^{(p)} = f_{\text{act}}'(u_{\lambda,\nu}^{(p)}) \sum_{\kappa=0}^{N_{\nu+1}-1} \delta_{\kappa,\nu+1}^{(p)} \, w_{\lambda,\nu,\kappa}. \tag{C.23}$$

Equations (C.20) and (C.23) define a recursive procedure for computing the $\delta_{\lambda,\nu}^{(p)}$ needed for computing the weight modification within (C.4). This procedure is known under the term backpropagation algorithm for artificial neural networks or multilayer perceptrons, respectively, in feed forward operation with non-linear processing units. The bias values can simply be computed using

$$\Delta_p \theta_{\lambda,\nu} = -\eta \delta_{\lambda,\nu}^{(p)}. \tag{C.24}$$

A toolkit to use this and other training algorithms is provided by [Uni 01].

D

Stability Check

To check whether an all-pole filter is stable we have to analyze if all poles are within the unit circle. So we can state the following constraint for the all-pole filter $\frac{1}{A(z)}$

$$|z| \leq 1 \left\{ z \in \mathbb{C} \mid A(z) = 0 \right\}. \tag{D.1}$$

There exist a lot of different analytical and numerical approaches to compute or approximate the zeros of $A(z)$. Exemplarily we will introduce a method that transforms the problem of finding the zeros of $A(z)$ to the search for the eigenvalues of a matrix with special properties depending on $A(z)$.

Let us first recall some basic considerations. The characteristic polynomial $\rho(t)$ of a matrix $\mathbf{P}$ is defined as

$$\rho(t) = \det\left(t\mathbf{I} - \mathbf{P}\right). \tag{D.2}$$

where $\mathbf{I}$ denotes the identity matrix. The eigenvalues λ of the matrix $\mathbf{P}$ are the roots of the characteristic polynomial

$$\rho(\lambda) = 0. \tag{D.3}$$

Given a normed polynomial or also monic polynomial $p(t)$ of order n which is characterized by $p_n = 1$

$$p(t) = p_0 + p_1 t + p_2 t^2 \ldots + p_{n-1} t^{n-1} + t^n, \tag{D.4}$$

we can define the so-called *companion matrix* $\mathbf{C}(p)$, which is a $n \times n$ matrix, as

$$\mathbf{C}(p) = \begin{bmatrix} 0 & 0 & \ldots & 0 & -p_0 \\ 1 & 0 & \ldots & 0 & -p_1 \\ 0 & 1 & \ldots & 0 & -p_2 \\ \vdots & \vdots & \ddots & \vdots & \vdots \\ 0 & 0 & \ldots & 1 & -p_{n-1} \end{bmatrix}. \tag{D.5}$$

This matrix has some interesting properties. The most interesting property concerning the usage for finding the roots of a polynomial is that the characteristic polynomial of this companion matrix equals the polynomial the companion matrix is based on

$$\det\left(t\mathbf{I} - \mathbf{C}(p)\right) = p(t).\tag{D.6}$$

Consequence of this property is that the task of finding the roots of $p(t)$ is equal to finding the eigenvalues λ of the corresponding companion matrix $\mathbf{C}(p)$. For a non-normalized polynomial $p(t)$ the companion matrix can easily be computed as

$$\mathbf{C}(p) = \begin{bmatrix} 0 & 0 & \cdots & 0 & -p_0/p_n \\ 1 & 0 & \cdots & 0 & -p_1/p_n \\ 0 & 1 & \cdots & 0 & -p_2/p_n \\ \vdots & \vdots & \ddots & \vdots & \vdots \\ 0 & 0 & \cdots & 1 & -p_{n-1}/p_n \end{bmatrix}.\tag{D.7}$$

Basically this is not a problem reduction but only a transform since the basic polynomial $p(t)$ equals the characteristic polynomial $\rho(t) = \det\left(t\mathbf{I} - \mathbf{C}(p)\right)$. The benefit is the possibility of using very efficient math libraries like *LAPACK* [Anderson 90] and the corresponding functions for finding the eigenvalues of a non-symmetric $n \times n$ matrix (e.g., *dgeev*).

After having computed the zeros of $A(z)$ we have to decide whether the resulting all-pole filter is stable and can be used as an estimate for the broadband spectral envelope or if we have to exchange the unstable all-pole filter by the estimate from the broadband codebook (see Sect. 5.6).

E

Feature Distributions

In this chapter of the appendix the various distributions of the cepstral coefficients that represent the spectral envelope and have been determined during the training phase are depicted. The figures show the distribution before a proper normalization has been performed. On the x-axis the value of each parameter is depicted and on the y-axis the corresponding relative frequency of this observation within the training data set. The coefficient c_0 is not depicted since $c_0 = 0$ holds for the narrowband as well as for the broadband coefficients. The distributions for each cepstral coefficient representing the narrowband spectral envelopes (telephone band limited) are presented in Figs. E.1–E.4.

The corresponding distributions of the cepstral coefficients representing the broadband spectral envelopes are presented in Figs. E.5–E.9.

As can be seen in the above presented figures, most of the distributions of the coefficients do look similar to a gaussian distribution. Another observation is that with increasing index of the coefficients the variance of the values of the coefficients decreases. This has been exploited to decrease the importance of deviations in the cepstral coefficients with a higher index during the training of the codebooks by the normalization presented in Sect. 5.3.1.

Another feature that has been investigated for using additional features to the parametric representations of the spectral envelope is the ACI feature which corresponds to the pitch frequency (see Sect. 3.3.3). The evaluation of this scalar feature makes sense only when there are voiced utterances. This can be assured by observing the ACM value (distribution presented below) which provides information that correlates with the voicedness of the utterance and the confidence of the pitch determination algorithm. This feature has also been used in a codebook approach presented in Sect. 5.3.2. The distribution of ACI in the training data set is depicted in Fig. E.10.

The next feature that has been investigated for being able to better separate between voiced and unvoiced utterances is the ACM value (see Sect. 3.3.3) that is calculated on basis of the maximum of the autocorrelation function. The distribution of this scalar parameter of the training data set has been determined as well and is depicted in Fig. E.11. The last additional feature

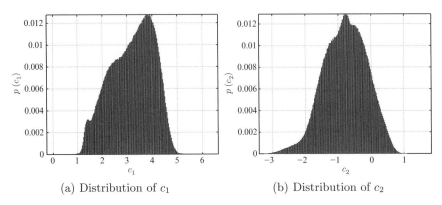

(a) Distribution of c_1 (b) Distribution of c_2

Fig. E.1. Distributions of the narrowband cepstral coefficients with index 1–2

that has been investigated for being included in a possible codebook approach is the highpass to lowpass energy ratio as discussed in Sect. 3.3.6. This feature serves as well as the ACM feature as an indicator for voicedness. The distribution of this feature in the training data set is depicted in Fig. E.12.

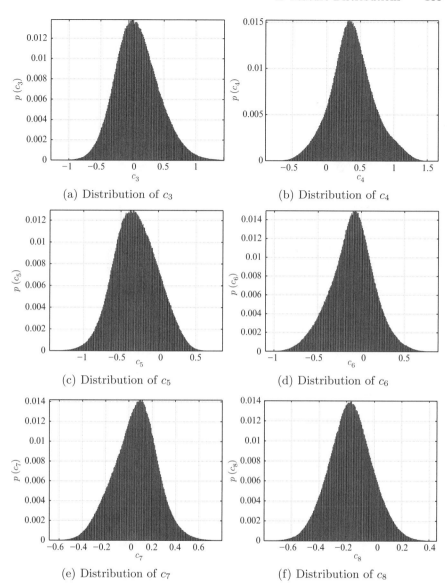

(a) Distribution of c_3

(b) Distribution of c_4

(c) Distribution of c_5

(d) Distribution of c_6

(e) Distribution of c_7

(f) Distribution of c_8

Fig. E.2. Distributions of the narrowband cepstral coefficients with index 3–8

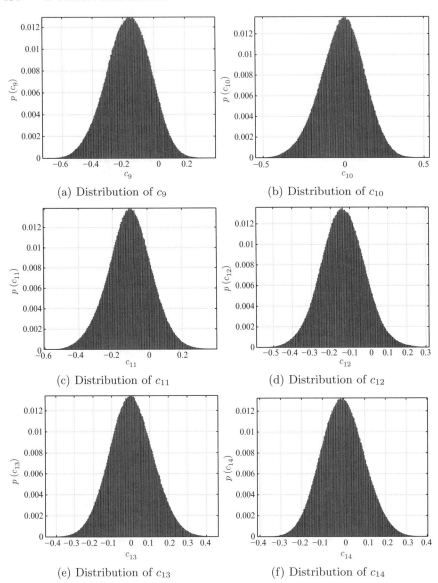

(a) Distribution of c_9

(b) Distribution of c_{10}

(c) Distribution of c_{11}

(d) Distribution of c_{12}

(e) Distribution of c_{13}

(f) Distribution of c_{14}

Fig. E.3. Distributions of the narrowband cepstral coefficients with index 9–14

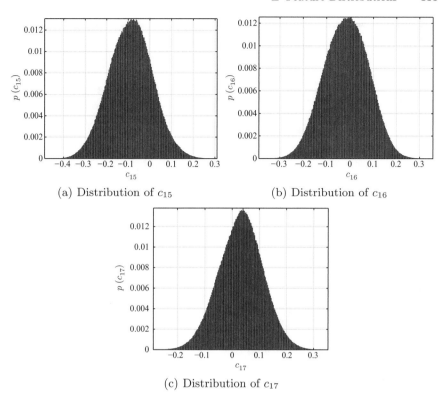

(a) Distribution of c_{15}

(b) Distribution of c_{16}

(c) Distribution of c_{17}

Fig. E.4. Distributions of the narrowband cepstral coefficients with index 15–17

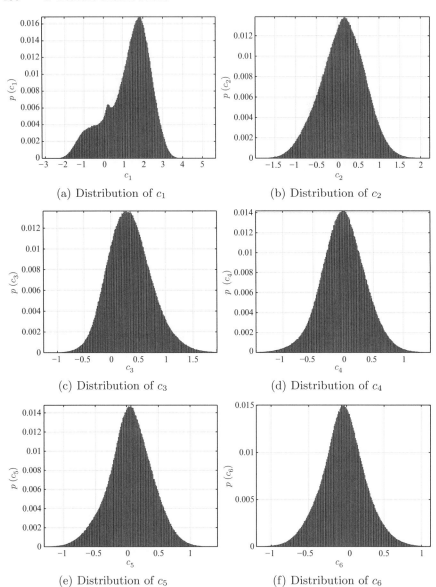

(a) Distribution of c_1

(b) Distribution of c_2

(c) Distribution of c_3

(d) Distribution of c_4

(e) Distribution of c_5

(f) Distribution of c_6

Fig. E.5. Distributions of the broadband cepstral coefficients with index 1–6

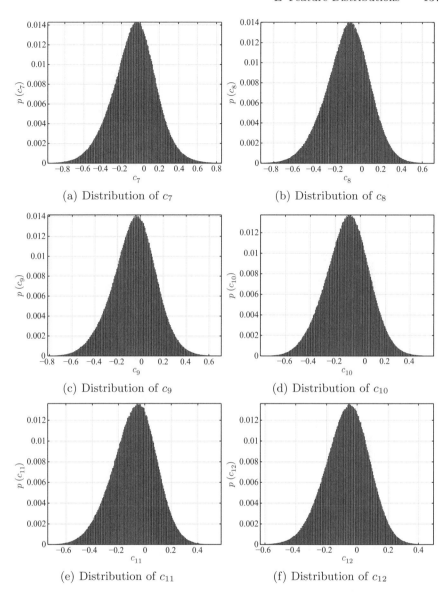

(a) Distribution of c_7

(b) Distribution of c_8

(c) Distribution of c_9

(d) Distribution of c_{10}

(e) Distribution of c_{11}

(f) Distribution of c_{12}

Fig. E.6. Distributions of the broadband cepstral coefficients with index 7–12

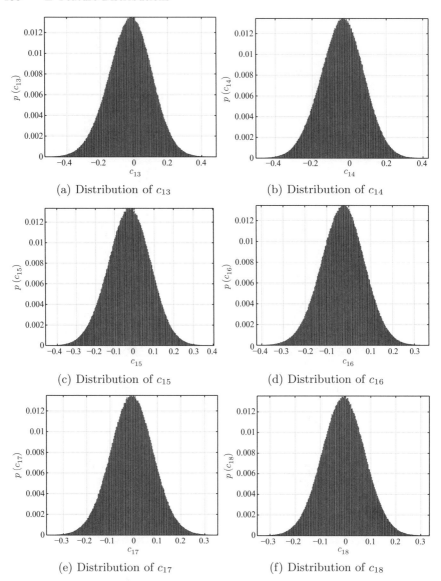

(a) Distribution of c_{13} (b) Distribution of c_{14}

(c) Distribution of c_{15} (d) Distribution of c_{16}

(e) Distribution of c_{17} (f) Distribution of c_{18}

Fig. E.7. Distributions of the broadband cepstral coefficients with index 13–18

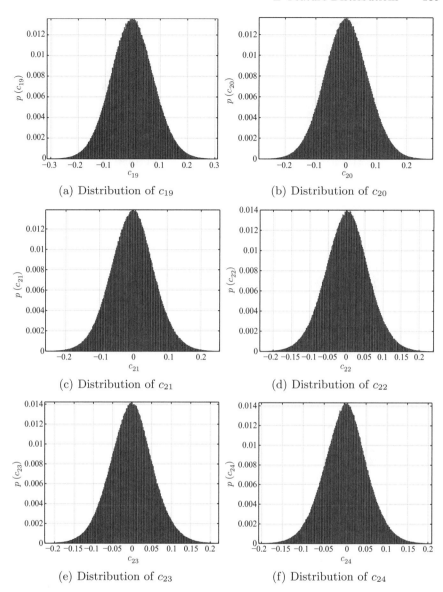

(a) Distribution of c_{19}

(b) Distribution of c_{20}

(c) Distribution of c_{21}

(d) Distribution of c_{22}

(e) Distribution of c_{23}

(f) Distribution of c_{24}

Fig. E.8. Distributions of the broadband cepstral coefficients with index 19–24

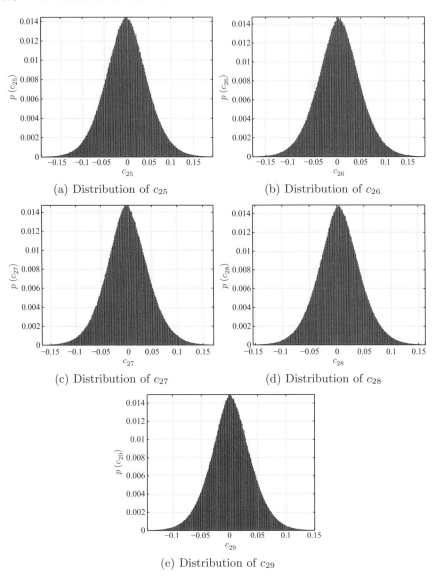

(a) Distribution of c_{25}

(b) Distribution of c_{26}

(c) Distribution of c_{27}

(d) Distribution of c_{28}

(e) Distribution of c_{29}

Fig. E.9. Distributions of the broadband cepstral coefficients with index 25–29

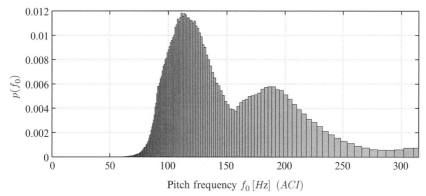

(a) Male and female speakers evaluated together.

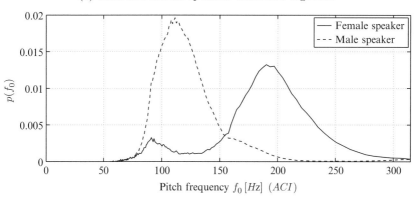

(b) Male and female speakers evaluated separately.

Fig. E.10. Distribution of the ACI values (pitch frequency) in Hz, evaluated for an ACM value (confidence/voicedness) >0.7. Note that the amount of frames of male and female speakers in the training data set having an ACM value >0.7 is not equal, causing the overall distribution not being equal to the scaled sum of the separate distributions. Also note that in (**a**) the width of the bars is increasing since $\propto \frac{f_\mathrm{s}}{\mathrm{tap}}$

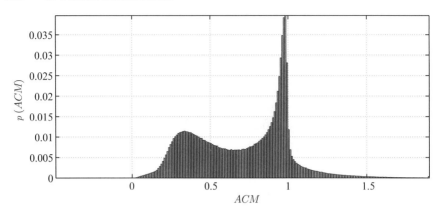

Fig. E.11. Distribution of ACM (voicedness/confidence)

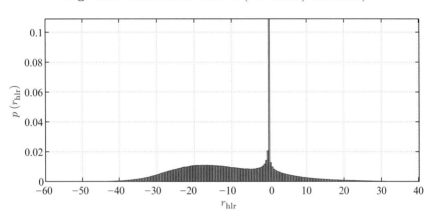

Fig. E.12. Distribution of r_{hlr} (voicedness)

F

Evaluation Results

In this chapter of the appendix the results of the evaluation discussed in Sect. 6.2.1 using other distance measures as well as only subband evaluations are presented. The setup for the various methods mentioned can be looked up in Table 6.3 on page 108 as well as on page 112. Tables F.1–F.3 use the Itakura–Saito distance. Tables F.4–F.6 use the log spectral deviation. Table F.7–F.8 use the "spectral distortion measure". The evaluation over the entire band can be found in Table 6.4 on page 110. Tables F.9–F.16 show the evaluation for the approaches with linear mapping and Tables F.17–F.24 show the various results for the combined approach using the codebook as a classification stage before applying the linear mapping.

Table F.1. Itakura–Saito distance measure evaluated for different methods and different codebook sizes over the full frequency range

Performances measured with d_{IS}					
Codebook size	Method A	Method B	Method C	Method D	Method E
2	0.7227	0.6477	0.8263	0.6500	0.7454
4	0.6678	0.5871	0.7475	0.5768	0.6127
8	0.6382	0.5614	0.6979	0.5451	0.5823
16	0.6532	0.5698	0.6607	0.5132	0.5979
32	0.6401	0.6318	0.5823	0.4942	0.6393
64	0.6231	0.5752	0.5974	0.4769	0.6901
128	0.6224	0.6186	0.5891	0.4745	0.6514
256	0.5892	0.6274	0.6079	0.4773	0.6762

Table F.2. Itakura-Saito distance measure evaluated for different methods and different codebook sizes over the upper frequency range

Codebook size	Performances measured with d_{IS}				
	Method A	Method B	Method C	Method D	Method E
2	1.2995	1.0929	1.5574	1.0795	1.4283
4	1.2723	1.0527	1.4461	1.0051	1.1350
8	1.2754	1.0734	1.3195	0.9884	1.1294
16	1.3588	1.1481	1.2839	0.9147	1.2160
32	1.3676	1.3059	1.1220	0.8830	1.3191
64	1.3477	1.1869	1.1964	0.8491	1.4184
128	1.3693	1.2868	1.1864	0.8623	1.3883
256	1.2884	1.3167	1.2418	0.8814	1.4175

Table F.3. Itakura-Saito distance measure evaluated for different methods and different codebook sizes over the lower frequency range

Codebook size	Performances measured with d_{IS}				
	Method A	Method B	Method C	Method D	Method E
2	0.4665	0.4569	0.4412	0.7398	0.3874
4	0.4221	0.3869	0.4084	0.6987	0.3376
8	0.3391	0.3118	0.3333	0.6863	0.3062
16	0.2800	0.2573	0.3149	0.8880	0.2698
32	0.2497	0.2587	0.2952	0.9859	0.2523
64	0.2326	0.2362	0.2586	1.0993	0.2701
128	0.2181	0.2482	0.2427	1.1391	0.2404
256	0.2092	0.2541	0.2326	1.1994	0.2594

Table F.4. Log spectral distance measure evaluated for different methods and different codebook sizes over the full frequency range

Performances measured with d_{LSD}					
Codebook size	Method A	Method B	Method C	Method D	Method E
2	7.8446	8.1099	8.1892	8.0518	7.5111
4	7.1928	7.2152	7.5578	7.2148	7.0467
8	6.6674	6.6289	7.3738	6.6753	6.5452
16	6.4088	6.3151	6.9622	6.2626	6.2085
32	6.1533	6.1896	6.6154	6.0327	6.1282
64	5.9829	5.9371	6.3563	5.8168	6.0724
128	5.8588	5.9069	6.1741	5.6537	5.8916
256	5.8057	5.7926	6.0793	5.5496	5.8475

Table F.5. Log spectral distance measure evaluated for different methods and different codebook sizes over the upper frequency range

Performances measured with d_{LSD}					
Codebook size	Method A	Method B	Method C	Method D	Method E
2	9.8073	10.2352	10.3037	10.0670	9.5521
4	9.1496	9.1878	9.5019	9.2397	9.0603
8	8.6485	8.6630	9.2409	8.7368	8.6672
16	8.5464	8.4829	8.8946	8.2925	8.3256
32	8.2954	8.3871	8.5435	8.0736	8.4032
64	8.1893	8.1788	8.3812	7.8965	8.4207
128	8.1645	8.2028	8.2509	7.7921	8.2904
256	8.2163	8.1784	8.2360	7.7707	8.2641

Table F.6. Log spectral distance measure evaluated for different methods and different codebook sizes over the lower frequency range

Performances measured with d_{LSD}					
Codebook size	Method A	Method B	Method C	Method D	Method E
2	6.4047	6.8493	6.8569	7.5062	6.1722
4	5.6448	5.7818	6.4383	6.6952	5.8019
8	5.2012	5.3079	6.0396	6.3381	5.2453
16	4.7867	4.9475	5.6109	6.1567	4.9860
32	4.6603	4.8844	5.3550	6.0875	4.7864
64	4.4610	4.6597	5.0165	6.0662	4.7204
128	4.2799	4.6454	4.8396	5.9880	4.5550
256	4.1505	4.5582	4.7271	5.9588	4.5660

Table F.7. Spectral distance measure (SDM) evaluated for different methods and different codebook sizes over the upper frequency range

Codebook size	Performances measured with d_{SDM}				
	Method A	Method B	Method C	Method D	Method E
2	17.5187	22.1931	22.7413	19.7906	12.6104
4	11.0606	12.5472	13.5756	12.6869	11.2442
8	8.2515	9.0745	11.9735	9.9412	9.5582
16	7.6371	8.1159	11.2724	8.6664	7.6053
32	7.4180	7.6655	9.3405	8.0834	7.6701
64	7.3560	7.4489	7.9216	7.7795	7.5042
128	7.1805	7.2549	7.6248	7.3114	7.1954
256	8.0361	7.0781	7.3776	7.1471	7.0364

Table F.8. Spectral distance measure (SDM) evaluated for different methods and different codebook sizes over the lower frequency range

Codebook size	Performances measured with d_{SDM}				
	Method A	Method B	Method C	Method D	Method E
2	0.7148	0.8964	0.9540	0.7398	0.7797
4	0.4932	0.5995	0.7816	0.5063	0.7102
8	0.4368	0.5065	0.7106	0.4371	0.5168
16	0.3632	0.4410	0.5841	0.3758	0.4625
32	0.3789	0.4245	0.5421	0.3779	0.4242
64	0.3589	0.4080	0.4853	0.3705	0.3996
128	0.3443	0.3928	0.4672	0.3618	0.3913
256	0.3268	0.3824	0.4522	0.3552	0.3764

Table F.9. Itakura-Saito distance measure evaluated for different methods and different codebook sizes over the full frequency range

Codebook size	Performances measured with d_{IS}		
	Method A + LM	Method B + LM	Method D + LM
2	0.4655	0.4661	0.5192
4	0.4401	0.4383	0.4503
8	0.4132	0.4147	0.4264
16	0.4069	0.4145	0.4197
32	0.3953	0.4120	0.3940
64	0.3939	0.4087	0.3828
128	0.3873	0.4152	0.3735
256	0.3803	0.4178	0.3686

Table F.10. Itakura-Saito distance measure evaluated for different methods and different codebook sizes over the upper frequency range

Codebook size	Performances measured with d_{IS}		
	Method A + LM	Method B + LM	Method D + LM
2	0.9286	0.9121	0.9230
4	0.8968	0.8757	0.8362
8	0.8490	0.8450	0.7958
16	0.8423	0.8596	0.7881
32	0.8216	0.8605	0.7328
64	0.8244	0.8521	0.7107
128	0.8123	0.8650	0.6949
256	0.7892	0.8699	0.6858

Table F.11. Itakura-Saito distance measure evaluated for different methods and different codebook sizes over the lower frequency range

Codebook size	Performances measured with d_{IS}		
	Method A + LM	Method B + LM	Method D + LM
2	0.2535	0.2601	0.5100
4	0.2353	0.2361	0.5552
8	0.2222	0.2231	0.5512
16	0.2115	0.2143	0.6635
32	0.2056	0.2121	0.7111
64	0.2092	0.2074	0.7605
128	0.1961	0.2102	0.7962
256	0.1926	0.2139	0.8404

Table F.12. Log spectral distance measure evaluated for different methods and different codebook sizes over the full frequency range

Codebook size	Performances measured with d_{LSD}		
	Method A + LM	Method B + LM	Method D + LM
2	5.5083	5.6363	6.9879
4	5.2307	5.3087	6.2077
8	5.0211	5.0574	5.8970
16	4.9389	4.9670	5.6382
32	4.8450	4.8655	5.4176
64	4.7936	4.7842	5.2678
128	4.7330	4.7480	5.1313
256	4.7605	4.7032	5.0331

Table F.13. Log spectral distance measure evaluated for different methods and different codebook sizes over the upper frequency range

Codebook size	Performances measured with d_{LSD}		
	Method A + LM	Method B + LM	Method D + LM
2	7.8502	7.9331	9.0828
4	7.5101	7.5319	8.3826
8	7.2821	7.2890	7.9895
16	7.1997	7.2447	7.7735
32	7.0920	7.1415	7.5229
64	7.0487	7.0487	7.3830
128	6.9691	7.0104	7.2583
256	7.0193	6.9627	7.1785

Table F.14. Log spectral distance measure evaluated for different methods and different codebook sizes over the lower frequency range

Codebook size	Performances measured with d_{LSD}		
	Method A + LM	Method B + LM	Method D + LM
2	4.5268	4.7239	6.4171
4	4.3218	4.4196	6.0651
8	4.1310	4.2108	5.8579
16	4.0286	4.1138	5.7184
32	3.9614	4.0387	5.6379
64	3.9589	3.9588	5.6096
128	3.8251	3.9440	5.5505
256	3.7703	3.9189	5.5140

Table F.15. Spectral distance measure (SDM) evaluated for different methods and different codebook sizes over the upper frequency range

Performances measured with d_{SDM}			
Codebook size	Method A + LM	Method B + LM	Method D + LM
2	7.3989	7.9004	13.0465
4	5.8659	6.1388	9.5745
8	5.2196	5.1910	7.7692
16	4.9978	5.0737	7.3202
32	4.8081	4.8134	6.7373
64	4.7446	4.6181	6.4987
128	4.5746	4.4656	6.1554
256	5.2019	4.3851	5.9602

Table F.16. Spectral distance measure (SDM) evaluated for different methods and different codebook sizes over the lower frequency range

Performances measured with d_{SDM}			
Codebook size	Method A + LM	Method B + LM	Method D + LM
2	0.3444	0.3933	0.5106
4	0.3075	0.3316	0.4119
8	0.2800	0.3009	0.3626
16	0.2701	0.2893	0.3268
32	0.2646	0.2764	0.3148
64	0.2558	0.2643	0.3084
128	0.2486	0.2572	0.3024
256	0.2418	0.2538	0.2938

Table F.17. Itakura-Saito distance measure evaluated for different methods and different codebook sizes over the full frequency range

Performances measured with d_{IS}				
Codebook size	Method I	Method II	Method III	Method IV
2	0.4768	0.4925	0.5119	0.4806
4	0.5102	0.5403	0.5867	0.5015
8	0.4688	0.5620	0.6046	0.4513
16	0.5057	0.6550	0.7374	0.5372
32	0.5447	0.8159	0.8312	0.6380
64	0.5670	0.8215	0.8767	0.5662
128	0.5486	0.8133	0.8850	0.5201
256	0.6516	0.9906	1.0743	0.5929

Table F.18. Itakura-Saito distance measure evaluated for different methods and different codebook sizes over the upper frequency range

Performances measured with d_{IS}				
Codebook size	Method I	Method II	Method III	Method IV
2	0.7377	0.7684	0.8188	0.7579
4	0.9000	0.9679	1.0891	0.8728
8	0.8613	1.0945	1.2104	0.7996
16	1.0111	1.3801	1.6214	1.0821
32	1.1345	1.7377	1.8697	1.2324
64	1.2142	1.7962	2.0011	1.1271
128	1.1788	1.7958	2.0583	1.0332
256	1.4082	2.1694	2.4615	1.1831

Table F.19. Itakura-Saito distance measure evaluated for different methods and different codebook sizes over the lower frequency range

Performances measured with d_{IS}				
Codebook size	Method I	Method II	Method III	Method IV
2	0.6896	0.6997	0.6993	0.6746
4	0.5845	0.5524	0.5637	0.5709
8	0.5076	0.4848	0.5097	0.5032
16	0.4745	0.4928	0.5152	0.4787
32	0.4665	0.5244	0.5603	0.4764
64	0.4593	0.5336	0.5522	0.4915
128	0.4570	0.5207	0.5374	0.4730
256	0.5171	0.6347	0.6858	0.5051

Table F.20. Log spectral distance measure evaluated for different methods and different codebook sizes over the full frequency range

Performances measured with d_{LSD}				
Codebook size	Method I	Method II	Method III	Method IV
2	6.9441	6.9086	6.9233	6.8716
4	6.7107	6.6541	6.7636	6.6230
8	6.1356	6.1938	6.2483	6.0958
16	5.8335	6.0361	6.1333	5.9154
32	5.7085	6.1527	6.0832	6.0282
64	5.5908	6.0147	6.0384	5.7016
128	5.4518	5.8583	5.8900	5.4635
256	5.5185	6.0609	6.1189	5.4975

Table F.21. Log spectral distance measure evaluated for different methods and different codebook sizes over the upper frequency range

Performances measured with d_{LSD}				
Codebook size	Method I	Method II	Method III	Method IV
2	8.0403	7.8683	7.8878	7.9349
4	8.2027	7.7716	7.9083	8.1304
8	7.6592	7.5284	7.5974	7.5807
16	7.5249	7.6454	7.8507	7.4602
32	7.5084	7.8975	7.9517	7.7839
64	7.4959	7.7983	8.0396	7.4283
128	7.3730	7.7323	7.9570	7.2183
256	7.4922	7.9929	8.2915	7.2502

Table F.22. Log spectral distance measure evaluated for different methods and different codebook sizes over the lower frequency range

Performances measured with d_{LSD}				
Codebook size	Method I	Method II	Method III	Method IV
2	9.0516	9.1204	9.1493	9.0160
4	8.1903	7.6579	7.5845	8.5974
8	7.8732	7.3769	7.3488	8.2009
16	7.2935	6.9938	6.9029	7.2269
32	7.1210	7.0839	6.8855	7.7428
64	6.9845	7.2931	6.7307	7.9444
128	7.0214	7.0369	6.7251	7.5777
256	7.0645	7.1682	6.9769	7.4989

Table F.23. Spectral distance measure (SDM) evaluated for different methods and different codebook sizes over the upper frequency range

Codebook size	Performances measured with d_{SDM}			
	Method I	Method II	Method III	Method IV
2	10.4940	8.1043	7.7688	9.8712
4	11.1673	7.1304	7.0816	10.2854
8	7.3724	4.8279	4.4973	7.3210
16	6.0340	4.4704	4.3240	5.3399
32	6.1025	4.3071	4.3967	5.9074
64	5.6892	3.7299	4.1119	5.5307
128	5.3165	3.5811	3.8132	5.0953
256	5.1048	3.3325	3.7145	4.7127

Table F.24. Spectral distance measure (SDM) evaluated for different methods and different codebook sizes over the lower frequency range

Codebook size	Performances measured with d_{SDM}			
	Method I	Method II	Method III	Method IV
2	2.0123	2.0495	2.0777	2.0150
4	1.4048	1.2241	1.1500	1.6770
8	1.3784	1.1885	1.1312	1.5930
16	1.0998	0.9561	0.8592	1.1063
32	1.0373	0.9555	0.7728	1.4101
64	0.9827	1.1289	0.7235	1.5938
128	1.1808	1.2190	0.9002	1.6649
256	1.1401	1.0316	0.8542	1.4444

G

Abbreviations and Symbols

G.1 Abbreviations

ACT	Activation
AMR	Adaptive multi-rate
AR	Autoregressive
ARMA	Auto-regressive moving average
BB	Broadband
BCB	Broadband codebook
BS	Band stop
BWE	Band width extension
CB	Codebook
CD	Compact disc
CEPS	Cepstral
DCT	Discrete cosine transform
DFT	Discrete Fourier transform
DTFT	Discrete time Fourier transform
DVD	Digital versatile disc
EFR	Enhanced full rate
EXT	Extended
FEAT	Feature
FFT	Fast Fourier transform
FIR	Finite impulse response
GI	Gradient index
GSM	Global system for mobile communication
HATS	Head and torso simulator
HLR	High to low frequency magnitude ratio
IDFT	Inverse discrete Fourier transform
IDTFT	Inverse discrete time Fourier transform
IIR	Infinite impulse response
ISDN	Integrated service digital network

LBG	Linde Buzo Gray
LK	Local Kurtosis
LP	Linear prediction
LPC	Linear predictive coding
LR	Likelihood ratio
LSD	Log spectral deviation/distortion
LSF	Line spectral frequencies
LSP	Line spectral pairs
MA	Moving average
MEL	Mel scale
MFCC	Mel frequency cepstral coefficient
MOD	Modified
MOST	Multimedia oriented system transfer
MRP	Mouth reference point
NB	Narrowband
NCB	Narrowband code book
NFE	Normalized frame energy
NORM	Normalized
NRFE	Normalized relative frame energy
PESQ	Perceptual evaluation of speech quality
PN	Pseudonoise
PSD	Power spectral density
SC	Spectral centroid
SDM	Spectral distortion measure
SNR	Signal to noise ratio
RFE	Relative frame energy
TB	Telephone band
TEL	Telephone
U	Unvoiced
V	Voiced

G.2 Important Symbols

α_i	ith LSF coefficient (angle)
$A\left(e^{j\Omega}\right)$	Inverse spectrum of the vocal tract
$A\left(e^{j\Omega}, n\right)$	Inverse spectrum of the vocal tract at time n
$A(z)$	Inverse frequency response of the vocal tract in the z-domain
$A(z, n)$	Inverse frequency response of the vocal tract in the z-domain at time n
ACI	Index of the maximum in the autocorrelation
ACM	Maximum of the autocorrelation

$\mathbf{a}$	Vector containing the LPC coefficients
a_i	ith coefficient of an all-pole model
$B(z)$	MA part of an ARMA-model in the z-domain
$\mathfrak{C}()$	Classification function
b_i	ith coefficient of an all-zero-model
c_i	ith cepstral coefficient
$c_{i,\mathrm{mfcc}}$	ith MFCC coefficient
$\delta\left(\Omega_0\right)$	Dirac defined in the frequency domain at frequency Ω_0
d_{ceps}	Cepstral distance
d_{I}	Itakura distance measure
d_{IS}	Itakura-Saito distance measure
d_{LR}	Likelihood ratio distance measure
d_{LSD}	Log spectral deviation
d_{SDM}	Spectral distortion measure
$\mathcal{E}\{\}$	Expectation operator
$E\left(e^{j\Omega_k}\right)$	Discrete spectrum of the excitation/predictor-error signal
E_n	Squared sum of a block of the predictor error signal at time n
$\hat{E}_{\mathrm{bb}}\left(e^{j\Omega}\right)$	Spectrum of the estimated broadband excitation signal
$E_{\mathrm{nb}}\left(e^{j\Omega}\right)$	Spectrum of the narrowband excitation signal
$e(n)$	Excitation/predictor-error signal at time n
$e_{n,\sigma}(m)$	Segment of predictor-error signal at time n including gain
$E_\sigma(z)$	Excitation/predictor-error signal in the z-domain including gain
$e_\sigma(n)$	Predictor-error signal at time n including gain
$\phi_{ss,n}(i,\ell)$	Short-time covariance function at time n
$\mathcal{F}\{\}$	DFT/DTFT
$\mathcal{F}^{-1}\{\}$	IDFT/IDTFT
$f_0(n)$	Fundamental frequency of a speech signal at time n
F_i	ith formant of a short-term speech spectrum
$g(n)$	Amount of voicing in the source-filter model at time n
g_{ext}	Gain factor for power adjustment of the extended excitation signal to the narrowband one
$H\left(e^{j\Omega}\right)$	Spectrum of the frequency response of the vocal tract
$H\left(e^{j\Omega},n\right)$	Spectrum of the frequency response of the vocal tract at time n
$H(z)$	Frequency response of the vocal tract in the z-domain
$H(z,n)$	Frequency response of the vocal tract in the z-domain at time n
$H_{\mathrm{tel}}(z)$	Frequency response of the telephone channel in the z-domain
$\Im\{\}$	Imaginary part
$\ln$	Logarithm to the base e
$\log_{10}$	Logarithm to the base 10
$P(z)$	Mirror polynomial in the z-domain
ΔP	Acoustic pressure difference measured in Pa
$Q(z)$	Anti-mirror polynomial in the z-domain
$\mathcal{R}()$	Ranking function
$\Re\{\}$	Real part

$\mathbf{R}_{ss,n}$	Matrix containing the autocorrelation coefficients at time n in Toeplitz structure
$r_{\mathrm{hlr}}(m)$	Highpass energy to lowpass energy ratio in block m
$r_{ss,n}(i)$	Auto correlation function of speech signal $s(n)$ at time n
$\mathbf{r}_{ss,n}$	Vector containing the autocorrelation coefficients at time n
$\sigma(n)$	Gain of the excitation signal at time n in the source-filter model
$S\left(e^{j\Omega}\right)$	Spectrum of the input speech signal
$S_{\mathrm{mel}}\left(e^{j\Omega_k}\right)$	Mel scaled discrete magnitude spectrum of input signal
$s(n)$	Input speech signal of the transmission system at time n
$s_{\mathrm{ext}}(n)$	Output speech signal of the BWE system at time n
$s_n(m)$	Segment of input speech signal at time n
$s_{\mathrm{tel}}(n)$	Input speech signal of the BWE system at time n
$\mathrm{sgn}\{\}$	Sign of a signal
$V()$	Voting function
$\bar{V}()$	Mean voting function
$W(m)$	Short-term energy of the input signal in block m
$\bar{W}(m)$	Smoothed (long-term) energy of the input signal in block m
$W_{\mathrm{NFE}}(m)$	Normalized frame energy in block m
$W_{\mathrm{noise}}(m)$	Estimate for the short-term noise energy in block m
$W_{\mathrm{NRFE}}(m)$	Normalized relative frame energy in block m
$W_{\mathrm{RFE}}(m)$	Relative frame energy in block m
$x_{s,\mathrm{gi}}(m)$	Gradient index of signal s at time m
$Z_s(m)$	Zero crossing rate of a segment of signal s starting at time m

List of Tables

List of Figures

References

[Agiomyrgiannakis 04]   Yannis Agiomyrgiannakis and Yannis Stylianou. *Combined Estimation/Coding of Highband Spectral Envelopes for Speech Spectrum Expansion*. In Proc. ICASSP, vol. 1, pages 469–472, Montreal, Canada, May 2004.

[Anderson 90]   E. Anderson, Z. Bai, J. Dongarra, A. Greenbaum, A. McKenney, J. Du Croz, S. Hammarling, J. Demmel, C. Bischof and D. Sorensen. *LAPACK: a portable linear algebra library for high-performance computers*. In Supercomputing '90: Proc. of the 1990 ACM/IEEE conference on Supercomputing, pages 2–11, Washington, DC, USA, 1990. IEEE Computer Society.

[Atal 82]   B. S. Atal. *Predictive Coding of Speech at Low Bit Rates*. IEEE Transactions on Communications, vol. COM-30, no. 4, pages 600–614, April 1982.

[Avendano 95]   C. Avendano, H. Hermansky and E. A. Wan. *Beyond Nyquist: Towards the Recovery of Broad-Bandwidth Speech from Narrow-Bandwidth Speech*. In Proc. Eurospeech, pages 165–168, Madrid, Spain, September 1995.

[Bansal 05]   Dhananjay Bansal, Bhiksha Raj and Paris Smaragdis. *Bandwidth Expansion of Narrowband Speech Using Non-Negative Matrix Factorization*. In Proc. Interspeech, pages 1505–1508, Lisbon, Portugal, September 2005.

[Bell 77]   Alexander Graham Bell. *The New Bell Telephone*. Scientific America, vol. 37, no. 1, 1877.

[Benesty 01]   J. Benesty, T. Gänsler, D. R. Morgan, M. M. Sondhi and S. L. Gay. *Advances in Network and Acoustic Echo Cancellation*. Digital Signal Processing. Springer, 2001.

168   References

[Boillot 05]   Marc Boillot and John Harris. *A Warped Bandwidth
                      Expansion Filter.* In Proc. ICASSP, vol. 1, pages 65–68,
                      Philadelphia, PA, USA, March 2005.

[Bronstein 85]   I. N. Bronstein and K. A. Semendjajew. *Taschenbuch
                      der Mathematik.* Verlag Harri Deutsch, 22nd edition, 1985.

[Carl 94a]   Holger Carl. *Untersuchung verschiedener Methoden der
                      Sprachcodierung und eine Anwendung zur Bandbreiten-
                      vergrößerung von Schmalbandsprachsignalen.* PhD thesis,
                      Ruhr-Universität Bochum, 1994.

[Carl 94b]   Holger Carl and Ulrich Heute. *Bandwidth Enhancement of
                      Narrow-Band Speech Signals.* In Proc. EUSIPCO, vol. 2,
                      pages 1178–1181, Edinburgh, Scotland, UK, September
                      1994.

[Chan 96]   C.-F. Chan and W.-K. Hui. *Wideband Re-synthesis of
                      Narrowband CELP-coded Speech Using Multiband Ex-
                      citation Model.* In Proc. ICSLP, vol. 1, pages 322–325,
                      Philadelphia, PA, USA, October 1996.

[Chan 97]   C.-F. Chan and W.-K. Hui. *Quality Enhancement of
                      Narrowband CELP-Coded Speech via Wideband Harmonic
                      Re-Synthesis.* In Proc. ICASSP, vol. 2, pages 1187–1190,
                      Munich, Germany, April 1997.

[Chen 04]   Guo Chen and Vijay Parsa. *HMM-Based Frequency
                      Bandwidth Extension for Speech Enhancement Using
                      Line Spectral Frequencies.* In Proc. ICASSP, vol. 1, pages
                      709–712, Montreal, Canada, May 2004.

[Cheng 94]   Y. M. Cheng, D. O'Shaughnessy and P. Mermelstein.
                      *Statistical Recovery of Wideband Speech from Narrowband
                      Speech.* IEEE Transactions on Speech and Audio Process-
                      ing, vol. 2, no. 4, pages 544–548, October 1994.

[Chennoukh 01]   S. Chennoukh, A. Gerrits, G. Miet and R. Sluijter.
                      *Speech Enhancement Via Frequency Bandwidth Exten-
                      sion Using Line Spectral Frequencies.* In Proc. ICASSP,
                      vol. 1, pages 665–668, Salt Lake City, UT, USA, May 2001.

[Croll 72]   M. G. Croll. *Sound Quality Improvement of Broadcast
                      Telephone Calls.* BBC Research Report RD 1972/26,
                      British Broadcasting Corporation, 1972.

[Cuperman 85]   V. Cuperman and A. Gersho. *Vector Predictive Coding of
                      Speech at 16 kbits/s.* IEEE Transactions on Communica-
                      tions, vol. COM-33, no. 7, pages 685–696, July 1985.

[Deller Jr. 00]   J. R. Deller Jr., J. H. L. Hansen and J. G. Proakis. *Discrete-Time Processing of Speech Signals.* IEEE Press, Piscataway, NJ, USA, 2000.

[El-Jaroudi 91]   Amro El-Jaroudi and John Makhoul. *Discrete All-Pole Modeling.* IEEE Transactions on Signal Processing, vol. 39, no. 2, pages 411–423, February 1991.

[Enbom 99]   N. Enbom and W. B. Kleijn. *Bandwidth Extension of Speech Based on Vector Quantization of the Mel Frequency Cepstral Coefficients.* In IEEE Workshop on Speech Coding, pages 171–173, Porvoo, Finland, June 1999.

[Eppinger 93]   B. Eppinger and E. Herter. *Sprachverarbeitung.* Carl Hanser Verlag, München Wien, 1993.

[Epps 98]   Julien Epps and W. Harvey Holmes. *Speech Enhancement Using STC-Based Bandwidth Extension.* In Proc. ICSLP, vol. 2, pages 519–522, Sydney, Australia, November/December 1998.

[Epps 99]   Julien Epps and W. Harvey Holmes. *A New Technique for Wideband Enhancement of Coded Narrowband Speech.* In IEEE Workshop on Speech Coding, pages 174–176, Porvoo, Finnland, June 1999.

[ETSI 00]   ETSI. *Digital Cellular Telecommunications System (Phase 2+); Enhanced Full Rate (EFR) Speech Transcoding.* EN 300 726 V8.0.1. ETSI, November 2000.

[Fuemmeler 01]   Jason A. Fuemmeler, Russell C. Hardie and William R. Gardner. *Techniques for the Regeneration of Wideband Speech from Narrowband Speech.* EURASIP J. Appl. Signal. Process., vol. 2001, no. 4, pages 266–274, 2001.

[Geiser 05]   Bernd Geiser, Peter Jax and Peter Vary. *Artificial Bandwidth Extension of Speech Supported by Watermark-Transmitted Side Information.* In Proc. Interspeech, pages 1497–1500, Lisbon, Portugal, September 2005.

[Gray 18]   H. Gray. *Grays Anatomy of the Human Body.* Lea & Febiger, Philadelphia, 20th edition, 1918.

[Gray 80]   R. M. Gray, A. Buzo, A. H. Gray and Y. Matsuyama. *Distortion Measures for Speech Processing.* IEEE Transactions on Acoustics, Speech and Signal Processing, vol. ASSP-28, no. 4, pages 367–376, August 1980.

[Gustafsson 01]   Harald Gustafsson, Ingvar Claesson and Ulf Lindgren. *Speech Bandwidth Extension*. In ICME. IEEE Computer Society, 2001.

[Hänsler 01]   Eberhard Hänsler. *Statistische Signale*. Springer, Berlin, 3rd edition, 2001.

[Hänsler 03]   Eberhard Hänsler and Gerhard Schmidt. *Single-Channel Acoustic Echo Cancellation*. In J. Benesty and Y. Huang, editors, Adaptive Signal Processing: Applications to Real-World Problems, chapter 3, pages 59–91. Springer, 2003.

[Hänsler 04]   Eberhard Hänsler and Gerhard Schmidt. *Acoustic Echo and Noise Control - A Practical Approach*. Adaptive and Learning Systems for Signal Processing, Communication and Control. Wiley, New York, 2004.

[Heide 98]   David A. Heide and George S. Kang. *Speech Enhancement for Bandlimited Speech*. In Proc. ICASSP, vol. 1, pages 393–396, Seattle, WA, USA, 1998.

[Hermansky 95]   H. Hermansky, E. A. Wan and C. Avendano. *Speech Enhancement Based on Temporal Processing*. In Proc. ICASSP, pages 405–408, Detroit, MI, 1995.

[Hosoki 02]   Mitsuhiro Hosoki, Takayuki Nagai and Akira Kurematsu. *Speech Signal Band Width Extension and Noise Removal Using Subband HMM*. In Proc. ICASSP, vol. 1, pages 245–248, Orlando, FL, USA, 2002.

[Hu 05]   Rongqiang Hu, Venkatesh Krishnan and David V. Anderson. *Speech Bandwidth Extension by Improved Codebook Mapping Towards Increased Phonetic Classification*. In Proc. Interspeech, pages 1501–1504, Lisbon, Portugal, September 2005.

[IPA 49]   IPA. *The principles of the International Phonetic Association, being a description of the International Phonetic Alphabet and the manner of using it, illustrated by texts in 51 languages*. University College, Department of Phonetics, London, 1949.

[Iser 03]   Bernd Iser and Gerhard Schmidt. *Neural Networks versus Codebooks in an Application for Bandwidth Extension of Speech Signals*. In Proc. Eurospeech, pages 565–568, Geneva, Switzerland, September 2003.

[Iser 05a]   Bernd Iser. *Bandwidth Extension of Speech Signals*. In 2nd Workshop on Wideband Speech Quality in Terminals and

Networks: Assessment and Prediction, Mainz, Germany, June 2005. ETSI.

[Iser 05b]   Bernd Iser and Gerhard Schmidt. *Bandwidth Extension of Telephony Speech.* EURASIP Newsletters, vol. 16, no. 2, pages 2–24, June 2005.

[Iser 06]   Bernd Iser and Gerhard Schmidt. *Bewertung verschiedener Methoden zur Erzeugung des Anregungssignals innerhalb eines Algorithmus zur Bandbreitenerweiterung.* In ITG-Fachtagung Sprachkommunikation, Kiel, Germany, April 2006.

[Iser 08]   Bernd Iser and Gerhard Schmidt. *Bandwidth Extension of Telephony Speech.* In Tulay Adali and Simon Haykin, editors, Advances in Adaptive Filtering. Wiley, New York, 2008.

[ITU 88a]   ITU. *7 kHz audio coding within 64 kbit/s.* Recommendation G.722. ITU-T, 1988.

[ITU 88b]   ITU. *General Performance Objectives Applicable to all Modern International Circuits and National Extension Circuits.* Recommendation G.151. ITU-T, 1988.

[ITU 01]   ITU. *Perceptual evaluation of speech quality (PESQ), an objective method for end-to-end speech quality assessment of narrowband telephone networks and speech codecs.* Recommendation P.862. ITU-T, Geneva, February 2001.

[Jax 00]   Peter Jax and Peter Vary. *Wideband Extension of Telephone Speech Using a Hidden Markov Model.* In IEEE Workshop on Speech Coding, pages 133–135, Delavan, Wisconsin, September 2000.

[Jax 02a]   Peter Jax. *Enhancement of Bandlimited Speech Signals: Algorithms and Theoretical Bounds.* PhD thesis, IND, RWTH Aachen, Aachen, Germany, 2002.

[Jax 02b]   Peter Jax and Peter Vary. *An Upper Bound on the Quality of Artificial Bandwidth Extension of Narrowband Speech Signals.* In Proc. ICASSP, vol. 1, pages 237–240, Orlando, FL, USA, 2002.

[Jax 03a]   Peter Jax and Peter Vary. *Artificial Bandwidth Extension of Speech Signals Using MMSE Estimation Based on a Hidden Markov Model.* In Proc. ICASSP, vol. 1, pages 680–683, Hong Kong, China, April 2003.

[Jax 03b]   Peter Jax and Peter Vary. *On artificial bandwidth exten-sion of telephone speech.* Signal Processing, vol. 83, no. 8, pages 1707–1719, August 2003.

[Jax 04a]   Peter Jax. *Bandwidth Extension for Speech.* In E. R. Larsen and R. M. Aarts, editors, Audio Bandwidth Extension, chapter 6, pages 171–235. Wiley, New York, 2004.

[Jax 04b]   Peter Jax and Peter Vary. *Feature Selection for Improved Bandwidth Extension of Speech Signals.* In Proc. ICASSP, vol. 1, pages 697–700, Montreal, Canada, May 2004.

[Jiménez-Fernàndez 98]   C. Jiménez-Fernàndez. *Wiedergewinnung von Breitband-sprache aus bandbegrenzter, codierter Sprache für ein Sprachspeichermodul.* Master's thesis, Universität Ulm, 1998.

[Kabal 03]   Peter Kabal. *Ill-Conditioning and Bandwidth Expansion in Linear Prediction of Speech.* In Proc. ICASSP, vol. 1, pages 824–827, Hong Kong, China, April 2003.

[Kallio 02]   Laura Kallio. *Artificial Bandwidth Expansion of Narrow-band Speech in Mobile Communication Systems.* Master's thesis, Helsinki University of Technology, Helsinki, Finland, December 2002.

[Kammeyer 92]   K. D. Kammeyer. *Nachrichtenübertragung.* B. G. Teubner, Stuttgart, german edition, 1992.

[Kleijn 95]   W. B. Kleijn and K. K. Paliwal, editors. *Speech Coding and Synthesis.* Elsevier Science, Amsterdam, Holland, 1st edition, 1995.

[Kornagel 01]   Ulrich Kornagel. *Spectral Widening of the Excitation Signal for Telephone-Band Speech Enhancement.* In Proc. International Workshop on Acoustic Echo and Noise Control (IWAENC), pages 215–218, Darmstadt, Germany, September 2001.

[Kornagel 02]   Ulrich Kornagel. *Spectral Widening of Telephone Speech Using an Extended Classification Approach.* In Proc. EUSIPCO, vol. 2, pages 339–342, Toulouse, France, September 2002.

[Kornagel 03]   Ulrich Kornagel. *Improved Artificial Low-Pass Extension of Telephone Speech.* In Proc. International Workshop on Acoustic Echo and Noise Control (IWAENC), pages 107–110, Kyoto, Japan, September 2003.

[Kornagel 06]   Ulrich Kornagel. *Techniques for Artificial Bandwidth Extension of Telephone Speech.* Signal Processing, vol. 86, no. 6, pages 1296–1306, June 2006.

[Korpiun 02]   C. A. Korpiun. *Physiologie der Sprache.* Script, Universität Essen, 2002.

[Krini 07]   M. Krini and G. Schmidt. *Model-Based Speech Enhancement.* In E. Hänsler and G. Schmidt, editors, Topics in Speech and Audio Processing in Adverse Environments. Springer, Berlin, prospectively published 2007.

[Laaksonen 05]   Laura Laaksonen, Juho Kontio and Paavo Alku. *Artificial Bandwidth Expansion Method to Improve Intelligibility and Quality of AMR-Coded Narrowband Speech.* In Proc. ICASSP, vol. 1, pages 809–812, Philadelphia, PA, USA, March 2005.

[LeBlanc 93]   W. P. LeBlanc, B. Bhattacharya, S. A. Mahmoud and V. Cuperman. *Efficient Search and Design Procedures for Robust Multi-Stage VQ of LPC Parameters for 4 kb/s Speech Coding.* IEEE Transactions on Speech and Audio Processing, vol. 1, no. 4, pages 373–385, October 1993.

[Linde 80]   Y. Linde, A. Buzo and R. M. Gray. *An Algorithm for Vector Quantizer Design.* IEEE Transactions on Communications, vol. COM-28, no. 1, pages 84–95, January 1980.

[Luenberger 69]   David G. Luenberger. *Optimization by Vector Space Methods.* Wiley, New York, 1969.

[Markel 76]   J. D. Markel and A. H. Gray. *Linear Prediction of Speech.* Springer, New York, 1976.

[Martin 01]   Rainer Martin. *Noise Power Spectral Density Estimation Based on Optimal Smoothing and Minimum Statistics.* IEEE Transactions on Speech and Audio Processing, vol. 9, no. 5, pages 504–512, July 2001.

[Mason 00]   M. Mason, D. Butler, S. Sridharan and V. Chandran. *Narrowband speech enhancement using fricative spreading and bandwidth extension.* In Proc. International Symposium on Intelligent Signal Processing and Communication Systems (ISPACS), pages 714–717, Honolulu, Hawaii, November 2000.

[Miet 00]   G. Miet, A. Gerrits and J.-C. Valière. *Low-Band Extension of Telephone-Band Speech.* In Proc. ICASSP, vol. 3, pages 1851–1854, Istanbul, Turkey, June 2000.

[Mitra 93]   S. K. Mitra and J. F. Kaiser. *Handbook For Digital Signal Processing*. Wiley, New York, 1993.

[Murthi 00]   Manohar N. Murthi and Bhaskar D. Rao. *All-Pole Modeling of Speech Based on the Minimum Variance Distortionless Response Spectrum*. IEEE Transactions on Speech and Audio Processing, vol. 8, no. 3, pages 221–239, May 2000.

[Nakatoh 97]   Yoshihisa Nakatoh, M. Tsushima and T. Norimatsu. *Generation of Broadband Speech from Narrowband Speech Using Piecewise Linear Mapping*. In Proc. Eurospeech, pages 1643–1646, Rhodes, Greece, September 1997.

[Nauck 96]   D. Nauck, F. Klawonn and R. Kruse. *Neuronale Netze und Fuzzy-Systeme*. Vieweg, 2nd edition, 1996.

[Nilsson 01]   M. Nilsson and W. Kleijn. *Avoiding Over-Estimation in Bandwidth Extension of Telephony Speech*. In Proc. ICASSP, vol. 2, pages 193–196, Salt Lake City, UT, USA, May 2001.

[Oppenheim 89]   A. V. Oppenheim and R. W. Schafer. *Discrete- Time Signal Processing*. Prentice-Hall, Englewood Cliffs, 1989.

[O'Shaughnessy 00]   D. O'Shaughnessy. *Speech Communications*. IEEE Press, 2nd edition, 2000.

[Paliwal 93]   K. K. Paliwal and B. S. Atal. *Efficient Vector Quantization of LPC Parameters at 24 Bits/Frame*. IEEE Transactions on Speech and Audio Processing, vol. 1, no. 1, pages 3–14, January 1993.

[Park 00]   K.-Y. Park and H. Kim. *Narrowband to Wideband Conversion of Speech Using GMM Based Transformation*. In Proc. ICASSP, vol. 3, pages 1843–1846, Istanbul, Turkey, June 2000.

[Parveen 04]   Shahla Parveen and Phil Green. *Speech Enhancement with Missing Data Techniques Using Recurrent Neural Networks*. In Proc. ICASSP, vol. 1, pages 733–736, Montreal, Canada, May 2004.

[Pulakka 06]   Hannu Pulakka, Laura Laaksonen and Paavo Alku. *Quality Improvement of Telephone Speech by Artificial Bandwidth Expansion — Listening Tests in Three Languages*. In Proc. Interspeech, pages 1419–1422, Pittsburgh, PA, USA, September 2006.

[Qian 03]   Yasheng Qian and Peter Kabal. *Dual-Mode Wideband Speech Recovery from Narrowband Speech*. In Proc. Eurospeech, pages 1433–1436, Geneva, Switzerland, 2003.

[Qian 04a]   Yasheng Qian and Peter Kabal. *Combining Equalization and Estimation for Bandwidth Extension of Narrowband Speech*. In Proc. ICASSP, vol. 1, pages 713–716, Montreal, Canada, May 2004.

[Qian 04b]   Yasheng Qian and Peter Kabal. *Highband Spectrum Envelope Estimation of Telephone Speech Using Hard/Soft-Classification*. In Proc. Interspeech ICSLP, vol. 4, pages 2717–2720, Jeju Island, Korea, October 2004.

[Rabiner 78]   L. R. Rabiner and R. W. Schafer. *Digital Processing of Speech Signals*. Signal Processing Series. Prentice-Hall, Englewood Cliffs, 1978.

[Rabiner 93]   L. R. Rabiner and B.-H. Juang. *Fundamentals of Speech Recognition*. Signal Processing Series. Prentice-Hall, Englewood Cliffs, 1993.

[Raza 02a]   Dar Ghulam Raza and Cheung-Fat Chan. *Enhancing Quality of CELP Coded Speech Via Wideband Extension by Using Voicing GMM Interpolation and HNM Re-Synthesis*. In Proc. ICASSP, vol. 1, pages 241–244, Orlando, FL, USA, May 2002.

[Raza 02b]   Dar Ghulam Raza and Cheung-Fat Chan. *Enhancing Quality of CELP Coded Speech Via Wideband Extension by Using Voicing GMM Interpolation and HNM Re-Synthesis*. In Proc. EUSIPCO, vol. 3, pages 309–312, Toulouse, France, September 2002.

[Raza 03]   D. G. Raza and C. F. Chan. *Quality Enhancement of CELP Coded Speech by Using an MFCC Based Gaussian Mixture Model*. In Proc. Eurospeech, pages 541–544, Geneva, Switzerland, September 2003.

[Rojas 96]   R. Rojas. *Neural Networks*. Springer, 1996.

[Schroeder 81]   M. R. Schroeder. *Direct (Nonrecursive) Relations Between Cepstrum and Predictor Coefficients*. IEEE Transactions on Acoustic Speech, and Signal Processing, vol. ASSP-29, no. 2, April 1981.

[Schüßler 94]   H. W. Schüßler. *Digitale Signalverarbeitung 1*. Springer, Berlin, 4th edition, 1994.

[Seltzer 05a]   Michael Seltzer and Alex Acero. *Training Wideband Acoustic Models using Mixed-Bandwidth Training Data via Feature Bandwidth Extension.* In Proc. ICASSP, vol. 1, pages 921–924, Philadelphia, PA, USA, March 2005.

[Seltzer 05b]   Michael L. Seltzer, Alex Acero and Jasha Droppo. *Robust Bandwidth Extension of Noise-Corrupted Narrowband Speech.* In Proc. Interspeech, pages 1509–1512, Lisbon, Portugal, September 2005.

[Shahina 06]   A. Shahina and B. Yegnanarayana. *Mapping Neural Networks for Bandwidth Extension of Narrowband Speech.* In Proc. Interspeech, pages 1435–1438, Pittsburgh, PA, USA, September 2006.

[Soong 93]   F. K. Soong and B.-H. Juang. *Optimal Quantization of LSP Parameters.* IEEE Transactions on Speech and Audio Processing, vol. 1, no. 1, pages 15–24, January 1993.

[StatSoft 02]   Inc.   StatSoft.   *Neural   Networks.* http://www.statsoftinc.com/textbook/stneunet.html, 1984-2002.

[Taori 00]   R. Taori, R. Sluijter and A. Gerrits. *HI-BIN: An Alternative Approach to Wideband Speech Coding.* In Proc. ICASSP, vol. 2, pages 1157–1160, Istanbul, Turkey, June 2000.

[Uncini 99]   Aurelio Uncini, Francesco Gobbi and Francesco Piazza. *Frequency Recovery of Narrow-Band Speech Using Adaptive Spline Neural Networks.* In Proc. ICASSP, Phoenix, AZ, USA, May 1999.

[Uni 01]   Universität Stuttgart, Institut für Parallele und Verteilte Systeme. *SNNS User Manual Version 4.2*, 2001.

[Unno 05]   Takahiro Unno and Alan McCree. *A Robust Narrowband to Wideband Extension System Featuring Enhanced Codebook Mapping.* In Proc. ICASSP, vol. 1, pages 805–808, Philadelphia, PA, USA, March 2005.

[Valin 00]   Jean-Marc Valin and Roch Lefebvre. *Bandwidth Extension of Narrowband Speech for Low Bit-Rate Wideband Coding.* In IEEE Workshop on Speech Coding, pages 130–132, Delavan, Wisconsin, September 2000.

[Vary 98]   P. Vary, U. Heute and W. Hess. *Digitale Sprachsignalverarbeitung.* B. G. Teubner, Stuttgart, 1998.

[Yao 05]   Sheng Yao and Cheung-Fat Chan. *Block-based Bandwidth Extension of Narrowband Speech Signal by using CDHMM.* In Proc. ICASSP, vol. 1, pages 793–796, Philadelphia, PA, USA, March 2005.

[Yao 06]   Sheng Yao and Cheung-Fat Chan. *Speech Bandwidth Enhancement Using State Space Speech Dynamics.* In Proc. ICASSP, vol. 1, pages 489–492, Toulouse, France, May 2006.

[Yasukawa 96a]   Hiroshi Yasukawa. *Adaptive Digital Filtering For Signal Reconstruction Using Spectrum Extrapolation.* In Proc. EUSIPCO, pages 991–994, Trieste, Italy, September 1996.

[Yasukawa 96b]   Hiroshi Yasukawa. *Restoration of Wide Band Signal from Telephone Speech using Linear Prediction Error Processing.* In Proc. ICSLP '96, vol. 2, pages 901–904, Philadelphia, PA, October 1996.

[Yasukawa 96c]   Hiroshi Yasukawa. *Signal Restoration of Broad Band Speech Using Nonlinear Processing.* In Proc. EUSIPCO, pages 987–990, Trieste, Italy, September 1996.

[Yegnanarayana 97]   B. Yegnanarayana, Carlos Avendano, Hynek Hermansky and P. Satyanarayana Murthy. *Processing Linear Prediction Residual for Speech Enhancement.* In Proc. Eurospeech, pages 1399–1402, Rhodes, Greece, September 1997.

[Yoshida 94]   Yuki Yoshida and Masanobu Abe. *An Algorithm to Reconstruct Wideband Speech from Narrowband Speech Based on Codebook Mapping.* In Proc. ICSLP, pages 1591–1594, Yokohama, Japan, September 1994.

[Zaykovskiy 04]   Dmitry Zaykovskiy. *On the Use of Neural Networks for Vocal Tract Transfer Function Estimation.* Master's thesis, University of Ulm, Ulm, 2004.

[Zaykovskiy 05]   Dmitry Zaykovskiy and Bernd Iser. *Comparison of Neural Networks and Linear Mapping in an Application for Bandwidth Extension.* In Proc. 10th International Conference on Speech and Computer (SPECOM), pages 695–698, Patras, Greece, October 2005.

[Zwicker 99]   E. Zwicker and H. Fastl. *Psychoacoustics.* Springer, Berlin, 2nd edition, 1999.

Index

(*continued from page ii*)

Advances in Industrial Engineering and Operations Research
Chan, Alan H.S.; Ao, Sio-Iong (Eds.)
2008, XXVIII, 500 p., Hardcover
ISBN: 978-0-387-74903-7, Vol. 5

Advances in Communication Systems and Electrical Engineering
Huang, Xu; Chen, Yuh-Shyan; Ao, Sio-Iong (Eds.)
2008, Approx. 700 p., Hardcover
ISBN: 978-0-387-74937-2, Vol. 4

Time-Domain Beamforming and Blind Source Separation
Bourgeois, J.; Minker, W.
2009, Approx. 200 p., Hardcover
ISBN: 978-0-387-68835-0, Vol. 3

Digital Noise Monitoring of Defect Origin
Aliev, T.
2007, XIV, 223 p. 15 illus., Hardcover
ISBN: 978-0-387-71753-1, Vol. 2

Multi-Carrier Spread Spectrum 2007
Plass, S.; Dammann, A.; Kaiser, S.; Fazel, K. (Eds.)
2007, X, 106 p., Hardcover
ISBN: 978-1-4020-6128-8, Vol. 1